The Architectural Press: London

Allan Hodgkinson
M.Eng., C.Eng., F.I.C.E., F.I.Struct.E., M.Cons.E.

Foundation Design

First published in 1986 by the Architectural Press Ltd,
9 Queen Anne's Gate, London SW1H 9BY

BRITISH LIBRARY CATALOGUING IN PUBLICATION DATA

Hodgkinson, Allan
   Foundation design.
   1.Foundations — Handbooks, manuals, etc.
   I.Title
   624.1'5   TA775

   ISBN 0-85139-837-5

Typeset by Stratatype in Baskerville on Linotron 202

Printed and bound in Great Britain by
Biddles Ltd, Guildford and King's Lynn

Contents

Structural engineering design is aided by appropriate sampling of materials of construction both before and after completion of the structure.

The development of soil mechanics in the last fifty years has added greatly to the knowledge of the behaviour of soils, particularly clay. Soils are not so regular as construction materials and sampling is less frequent, so that greater judgement and experience are required of the designer of foundations.

As with the author's previous work, the *AJ Handbook of Building Structure,* the emphasis is on practical matters and, while every architect may not aspire to more complicated designs, with the aid of this book he will be able to talk with more authority to his engineer.

At the time of writing the book there was considerable controversy in the philosophy of design, the trend being from allowable stress to limit state. It is the author's belief that any method of design which satisfies the Building Regulations should be acceptable and that to decide by law that there should be only one method of design is a totally unwarranted and retrograde step. Fortunately, no method has yet been devised by which partial factors for material can be applied in foundation or retaining wall design. The book is therefore based on the accepted practices of the last eighty years.

I am grateful to the following for permission to reproduce illustrations and tables: Bartholomew, Fig. 1.1; Building Research Establishment, Tables 4.1, 4.2, Fig. 4.6; GKN-Keller, Fig. 4.31, 4.32, 4.37, 4.39; Piggot Foundations – Ground Engineering, Fig. 3.33; Wembley Laboratories, Figs. 2.3, 2.4, 2.5, 2.8, 2.10, 2.12; Henry Boot Civil Engineering, Fig. 1.4; British Standards Institution, Tables 1.1, 1.2, 1.3; Institution of Structural Engineers, Fig. 2.5; Fairbriar Homes Ltd, Fig. 3.3. I am particularly grateful to my wife, who patiently typed the manuscript, and to Terry Roberts, who drew all the line illustrations.

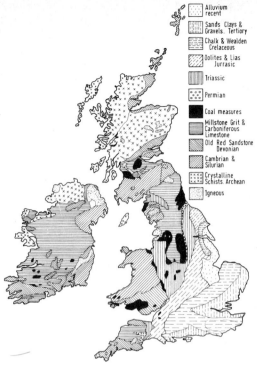

Chapter 1

Figure 1.1 Geological map of the British Isles.

Legend:
- Alluvium recent
- Sands Clays & Gravels. Tertiory
- Chalk & Wealden Cretaceous
- Oolites & Lias Jurrasic
- Triassic
- Permian
- Coal measures
- Millstone Grit & Carboniferous Limestone
- Old Red Sandstone Devonian
- Cambrian & Silurian
- Crystalline Schists. Archean
- Igneous

Introduction to Rocks and Soils

The earth's naturally occurring deposits are classified by engineers as soil or rock with an arbitrary division based on strength, related physical properties and use. The engineering properties are largely determined by the properties of the component grains such as particle size, shape and surface texture, structures in the deposits such as bedding and mechanical discontinuities, chemical composition and the packing and orientation of the grains. The disposition of rocks and soils in the United Kingdom is shown in Figure 1.1.

1.1   Rocks

Rocks are often thought of as some incompressible bed on which almost any building can be accommodated, but the manner in which the rock was formed and its subsequent deformation by the earth's movements determine its allowable bearing capacity.

Rocks are assessed geologically under three headings:

(a)   *Igneous*, such as basalts and granites, formed mainly by solidification from molten or liquid material.

(b)   *Sedimentary*, made from rock or mineral particles which have been transported or precipitated in water. They include shales, mudstones, sandstones, limestones, chalk and coal.

(c)   *Metamorphic*, which are crystalline rocks resulting from existing sedimentary or igneous rocks and are re-crystallised under the action of heat or pressure. They include gneisses, schists and slates.

The loadbearing capacity of all rocks is greatly reduced if they are decomposed, heavily shattered by earth movements or steeply dipping (Figures 1.2 and 1.3).

Figure 1.2   Anticlines and synclines (asymmetric and sometimes inverted) in limestone and shale, Loughshinny, Co. Dublin.

Figure 1.3 Dipping in limestone beds Salema, Algarve. Note the weathering of the softer layers and the many laminations.

Table 1.1 Presumed bearing values of rocks after CP 2004.
Note – ultimate bearing value is the pressure at which the foundation would fail in shear or continue to settle.
Presumed bearing value is the pressure with an adequate factor of safety ignoring factors such as degree of settlement, etc.
Allowable bearing pressure is that pressure which, having taken account of factors such as settlement, water table, etc., will provide an adequate factor of safety.

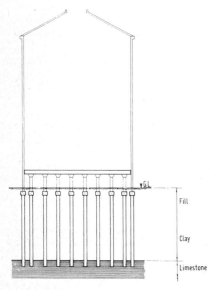

Figure 1.4 Foundation system of a sugar silo, Humberside.

Table 1.1, adapted from BS CP 2004, gives an indication of the presumed bearing pressures (hereafter referred to as PBP) which may be used in design. Where the rock is unfissured, triaxial compression tests using high lateral pressures, carried out on cores drilled from the rock will give a reasonable assessment of its strength. If the rock is fissured then no laboratory test can simulate the actual conditions. For example, for the foundations of a sugar silo on Humberside (Figure 1.4) a considerable depth of limestone was found under fill and soft clays at a depth of 8·5 m but it was layered with softer material in the bedding planes. Although the parent rock was assessed at a presumed bearing value of 700 tonnes per sq.m the working stress chosen was only 488 tonnes and this assumed a closing of the beds by about 20 mm in twenty years. In view of the susceptibility of the structure to differential movement both when full and empty the sugar-load platform was isolated from the outer wall which was carried on a ring beam, both structures being carried on end-bearing bored piles.

Rock type	Presumed bearing value kN/m²
Igneous, limestone and metamorphic rocks in massive form	10,000 or more
Sandstone unweathered medium to fine grain	7,000
Sandstone weathered	500–900
Schists and slates	3,000
Slightly weathered, closely jointed fine grained mudstone	5,000
Soft limestone	600
Keuper marl (unleached siltstone) unweathered	500–2,000
Keuper marl weathered	50–150
Keuper marl (mudstone) (varies relative to zone)	100–1,200
Chalk Grade 1	2,000
Grade 2	1,000
Grade 3	500
Grade 4	250
Grade 5	125

Thinly bedded limestones, sandstones, shales and heavily shattered rocks to be assessed after inspection

Chalk is considered in this rock group but it is not a material for the inexperienced to try to predict its PBP. It is graded in five classes, the upper being a hard rock, the lower almost the equivalent of a firm to soft clay. The cellular structure holds water but can be broken down easily and softened by frost, water or mechanical disturbance. Strength can be assessed by triaxial tests on cores, plate-bearing tests or penetration tests but it is essential that the chalk is seen in bulk before using these latter results to attribute a PBP.

1.2 Sands and Gravels

These are the principal non-cohesive soils. Shear strength is frictional and the structural properties depend mainly on the density and the distribution of size of the particles. Shear failure is unlikely except in very narrow shallow footings on loose sand near the water table and PBP is governed by settlement considerations, though in a building the settlement is probably complete by the time the building is topped out. Loose sand can settle and lose stability under shock or vibration and in the presence of water a 'quick sand' condition can develop. Apart from the fine sand, sand and gravels are not liable to frost heave. PBP is assessed from penetration tests and the considerations of density and particle size determined from sieve tests. Table 1.2, adapted from BS CP 2004, gives an indication of PBP.

Table 1.2 Presumed bearing values of cohesionless soils after CP 2004.

Cohesionless soils	N value	Presumed bearing value kN/m² for foundation widths	
		1 m	3 m
Very dense sand and gravels	>50	600	450
Dense sand and gravels	30–50	300–600	250–450
Medium dense sand and gravels	10–30	150–350	100–250
Loose sands and gravels	5–10	50–150	50–100

It is assumed that the water table is at a depth below the base of the foundation equal to the width of the foundation. Otherwise it must be reduced

The N values may be varied relative to the ground pressure at the depth of the test. See Chapter 2

1.3 Clays

Clay soils have very small voids or pores and tend to retain water in them. Normally they are relatively impermeable, settle slowly under load, and are cohesive, with negligible apparent frictional shear strength.

Because of pore water, they are prone to shrinkage and swelling and can creep down slopes. Partings or other materials and fissures in certain clays, and the presence of rootlets, can lead to general water seepage, softening and other weaknesses. Because of low shear strengths, rotational slips and excavation heave are possible. Strength and settlement are both criteria in foundation design. Clays frequently contain sulphates which attack Portland cement concrete and aggravate corrosion of buried ferrous metals.

According to the degree of pressure under which the original fine mud deposit was consolidated, the toughest clays become mudstone or shale. Mudstone may be massively bedded or laminated as a shale, which might contain silt. These materials are prone to considerable deterioration through weathering, and may soften in contact with water. Shales subject to prolonged pressure or heat are metamorphosed into true rock (hard shales or slates). Because of lamination, shales and slates are liable to slip in inclined strata.

Important clays are the stiff-to-hard boulder clays (which contain random stones and were deposited by glaciers), the 'shrinkable' stiff fissured clays (e.g. London, Lias, Weald and Oxford clays and those of the Woolwich and Reading beds), the soft alluvial and marine clays and the marls.

The stiff, fissured and boulder clays which have been subjected to great pressures during their geological history are known as 'over-consolidated' or 'pre-consolidated'; other clays are called 'normally consolidated'. The higher the degree of preconsolidation, the less compressible is the material and the lower the 'geological' reduction factor (μ) relating actual to laboratory consolidation. But some soft clays are very sensitive, i.e. they exhibit a large loss of shear strength if disturbed. (Sensitivity is the ratio of undisturbed to remoulded shear strength.)

Clays containing organic matter or interleaved with it are particularly compressible. Alluvial clays are normally consolidated; this gives them a marked increase in shear strength with depth compared with preconsolidated clays, but the top metre or so may be partially dried into a stiff surface crust. All clays are, of course, liable to softening by disturbance and water action and may disintegrate if allowed to dry out completely. Due to their cohesion firm clays will stand at steeply cut slopes for limited periods. Stability will eventually be impaired by weathering. 'Marl' normally describes calcareous clays, i.e. those containing calcium carbonate, heavily over-consolidated and very strong when unweathered, but they are often highly fissured and easily softened by water and disburbance. Claystones are hard concretions of clayey material cemented by calcium carbonate, often very large and occurring at certain levels in London and Oxford clays, etc.

Clays having similar characteristics to the UK stiff fissured clays are found in Montana and Saskatchewan in North America and in Trinidad, Denmark, West Germany and northern France.

Alluvial clays are found in eastern Canada, Norway and Sweden and, as in the case of the stiff fissured clays, precautions must be taken against seasonal movement and the action of vegetation roots.

Saline calcareous clays are widely distributed in the Near and Middle East and in Utah and Nevada. Formed by the deposition of clay minerals in saline water or water rich in lime, they are augmented by wind-blown sand and dust. The profile is comprised of a surface crust of about 2 m thick of hard to stiff desiccated clay overlying soft, moist clay. The surface crust is not softened to any appreciable extent by rain and has adequate strength to support light structures. Calcareous clays show marked volume changes with varying moisure content and in the countries bordering the Mediterranean, the soil movements extend to 5 m or more. Southern Iraq, with no marked difference in seasonal rainfall, does not have this problem.

Tropical black clays are found in specific parts of Sudan, Kenya, Zimbabwe, south-western States of America, India, Nigeria and Australia; unlike the red clays generally found in India, Africa, Hawaii, West Indies, South America and Far Eastern countries, they are very poor engineering materials in that they exhibit marked volume changes with changes in moisture content, and with poor drainage characteristics become impassable to construction traffic in the wet seasons. For example, in the Sudan, where seasonal swelling and shrinkage occur to a depth of 5 m, it may be necessary to pile even light buildings.

The red clays are largely residual soils from the physical and chemical weathering of igneous rocks, usually clays with a high PBP and low compressibility providing leaching does not occur. Table 1.3, adapted from BS CP 2004, gives an indication of PBP.

Table 1.3 Presumed bearing values of cohesive soils after CP 2004.

Cohesive soils	Cohesion	Presumed bearing value kN/m² for foundation widths	
		1 m	3 m
Hard boulder clays and fissured clays, strong weathered shales and weathered mudstones	>300	800	500
Very stiff boulder clay London clay, weathered Keuper marl	150–300	400–800	250–380
Stiff fissured clays	75–150	200–400	120–200
Firm normally consolidated clays	40–75	100–200	60–90
Soft normally consolidated alluvial clays	20–40	50–100	25–50

Assumes the foundations are at a minimum depth of 1 m below finished ground level

1.4 Silts and Peat

Properties of silts are intermediate between those of sands and clays, being both cohesive and frictional. Peat is a soft fibrous material formed by the decay of plants. Both materials are difficult foundation materials and are considered further in chapter 4.

General References

Smith, B., *Physical Geography for Schools*, A. & C. Black, 1917.

Smyth, F.G.H., *A Geology for Engineers*, Edward Arnold, 1945.

Nixon, I.K. and Skipp, B.D. 'Airfield construction on overseas soils', *Proceedings of ICE*, 1957.

Tomlinson, M.J., 'Airfield construction on overseas soil', *Proceedings of ICE*, 1957.

Clare, K.E. 'Airfield construction on overseas soils', *Proceedings of ICE*, 1957.

Higginbottom, J.E. 'The engineering geology of chalk', Symposium on Chalk in Earthworks and Foundations, ICE, 1966.

BS CP 2004: 1972 Code of Practice for foundations, BSI (under revision).

Skempton, A.W., *The Bearing Capacity of Clays*, Building Research Congress Division, 1, 1980.

Chapter 2

Site Investigation, Soil Mechanics and the Principles of Foundation Design

2.1 Site Investigation

Investigation of the site is best started by a walking round the site and studying existing information. In any tour of a site the following points may be noted:

(a) Indication of strata in banks, cuttings, ditches, nearby drainage excavations and quarry faces.

(b) Broken or stepped ground may give an indication of landslips or faults. Leaning trees or slopes or terracing of the surface steeper than 1 in 10 could be due to clay creeping down the slope.

(c) Polygonal surface cracking or crazing in very dry weather is an indication of a shrinkable clay. Larger, more or less parallel cracks may be caused by such things as landslips and mining subsidence. After a cold spell the effects of frost heave may be seen.

(d) In chalk or limestone, swallow-holes may be seen which have been caused by cavities in the underlying rock. Similar depressions or subsidence, or the unevenness of the adjacent road, may be a sign of mineral extraction.

(e) If an area is much lower lying than its surroundings it may contain soft alluvial deposits such as silt or peat, particularly so near rivers and some coastal regions. The possibility in the latter case of flooding must also be considered.

(f) Sea, rivers and springs are likely causes of immediate surface erosion.

(g) Water levels in nearby ditches and ponds may give some indication of the water table but this should be confirmed by the site investigation. Marshy vegetation and willows suggest a high water table.

(h) Sudden changes in the type of vegetation may be due to change in the subsoil conditions. Cracks or depressions near trees or shrubs may indicate a shrinkable clay.

(i)   Nearby buildings should be examined to see if they have cracked or deformed. If there are new buildings then the local authority are likely to be able to give a general impression of foundation conditions in the development.

(j)   Underground services or old foundations may be apparent from the surface features.

Having obtained a general impression of the site by the walk round, further work can then be carried out in the office or at the office of the local authority.

(a)   Establish what Ordnance Surveys are available and see how the site has been developed over the years. It may perhaps have been built on two or three times.

(b)   Consult regional geology books and seek the advice of site investigation contractors who may have knowledge of the area.

(c)   Carry out a very approximate assessment of the dead and superimposed loads to be carried and, on the basis of the architect's preferred support points in the building, assess whether individual foundations or a raft are a likely solution.

(d)   Having regard to the area of the site, the positive nature of the development on the site, i.e. the known positions of the buildings, the accessibility of plant to proposed boreholes (e.g. the site may be a city area with existing buildings which are later to be demolished), decide on the number of boreholes or trial pits required from the technical aspect.

(e)   On the basis of the building sizes (see below) decide the depth to which boreholes should be taken (trial pits are normally excavated by machine to not more than 4 m) and make an approximate estimate of the cost of the investigation. Discuss this with the client, relating the cost as a percentage of the total development cost, including the cost of the land. It is usually seen to be in the region of $\frac{1}{2} - 1\%$ and the client must be convinced by argument that this is an area of design where penny-pinching is not to his advantage.

With the knowledge of the site and the benefit of the further investigation, tenders can now be obtained from, say, three contractors, preferably members of the Association of Ground Investigation Specialists.

A typical investigation to tender for a small development is shown in Figure 2.1 with an approximate Bill of Quantities for the contractor to price and return. The contractor will also return a schedule of his laboratory testing costs.

Inevitably a site investigation is of a provisional nature. How positive it is depends upon the amount of information about the strata obtained before the investigation. It is not possible to establish how much testing will be required and this is usually expressed as a provisional sum in the inquiry.

It is preferable to proceed with the deepest borehole first and if the architect or engineer are present a rapid assessment can be made of the depth to which the

Figure 2.1 Typical invitation to tender and specification for a small site investigation.

SPECIFICATION FOR SITE INVESTIGATION

1. The Site is at the North end of Whitehorse Lane E17 in the London Borough of Withamstow and is bound on the South by existing factories, to the East by Whitehorse Lane, to the West by the River Lea and Reservoirs and to the North by the access road from Whitehorse Lane to the reservoirs.

2. The proposed development is indicated on the Architect's plan 2716/6 and will be typical portal frame construction on not more than a 6 m module. Plan 7830/2 shows the type of foundation loading which will occur assuming that ground beams are employed to carry perimeter brick, block and clad walls between the foundation bases or pile caps.

Ground slab loading will be 5 cwt/ square foot (25 KN/ square metre) and it is hoped to use a concrete slab floating within the perimeter shell with such excavation refilling or ground improvement as is appropriate.

3. Having regard to the marshes prevalent in the area, the Bill of Approximate Quantities has been written on the assumption that piling might be necessary and the quantities are based on 2 No. 20 m boreholes near to the river and 3 No. 10 m boreholes nearer to Whitehorse Lane.

On completion of the first 20 m borehole near the river, advice will be given with respect to the remainder of the depths to be bored.

Contact must be maintained with the Engineer's Office throughout the operation as it is essential that one of his staff is available at the Site for the boring of the bottom 10 m of the first borehole.

4. The work shall be carried out in accordance with BS 5930 and BS 1377. You shall assure yourselves that the type of rig you intend to employ can be moved across the site which is largely old garden allotments.

5. A PC Sum has been allowed for tests in the Laboratory to a programme to be agreed with the Engineer.

You are asked to submit your scale of test charges with your offer.

6. Will you please state the earliest date at which you could provide a rig on site and the time you would require to carry out the work outlined in the Approximate Bill of Quantities and the further time required to carry out the laboratory tests and prepare a Report.

	Description	Quantity	Unit	Rate £	p
1.	Carriage of plant and materials, men's fares and travelling time to and from the site		Lump sum		
2.	Move and set up rig at each boring position	5	No.		
3.	Boring in soft materials such as clay, silt, sand and gravel and mixtures of these with the exception of cobbles and boulders from surface to a depth not exceeding 5 m	25	Lin.m		
	To depth exceeding 5 m but not exceeding 10 m	25	Lin.m		
	To depth exceeding 15 m but not exceeding 20 m	20	Lin.m		
	Normal diameter of borehole between 130 and 200 mm				
4.	Extra over soft boring rate for overcoming obstructions such as concrete masonry, boulders, stones exceeding 100 mm diameter and strata requiring the use of chiselling techniques.		Per plant hour		
5.	Standing time for rig and crew		Per plant hour		
6.	Taking, packing and transporting to the laboratory of approximately 100 mm diameter undisturbed samples of cohesive materials	25	No.		
	Taking, packing and transporting to the laboratory of large disturbed samples of granular material	20	No.		
	Taking and transporting to the laboratory of groundwater samples	5	No.		
7.	Perform in situ standard penetration tests	20	No.		
8.	Soil testing to be agreed with Engineer	Provisional sum		250.00	
9.	Provide six copies of a comprehensive report comprising boring records, test results and discussion of site conditions		Sum		
10.	Provide any necessary barriers for the protection of the public during the works		Sum		
11.	Provide for water or other services to be used in the investigation		Sum		
12.	Clear up on completion of the work including filling in the boreholes		Sum		
13.	Contingency sum			100.00	
			TOTAL		

other boreholes need to be taken and the number and type of samples to be obtained. If there is no clear indication of the water table then a perforated pipe may be inserted in the borehole for future observation. Borehole logs are kept by the driller with his impression of the material encountered, but the description may well be amended by the soils engineer when related to the samples. If the loads and type of foundation are given to the laboratory, the soils engineer will probably recommend a programme of testing getting the most out of the provisional sum allowed or suggesting an increase or reduction in the sum.

2.2   The Mechanics of the Investigation

Trial pits

Trial pits are acceptable when the building places little load on the ground, e.g. two-storey housing or where the experience of the area in question has revealed no difficulties such as a much weaker stratum at a lower level. It is normal procedure to investigate to a depth of 1·5 times the width of the foundation, taking into account also the proximity of foundations (Figure 2.2), the object being to take the excavation to a depth where the load intensity on the soil is only a fifth of that taken over the area of the foundation. Excavation can be by hand but preferably only when access is difficult as this is a very expensive operation. As mentioned earlier, a variety of mechanical excavators are available, the biggest of which will be able to dig to a depth of 5 m. Whether the excavation stands long enough to be examined or whether it is safe to go down 4 m to examine will depend on the materials found. Clays and gravels will usually stand vertically to this depth provided there is no ground water problem. Sands are a little more doubtful, expecially if loose, irrespective of the water. Having the excavator bucket in the hole during examination is some safeguard but if in doubt a few sheeters and props are desirable. The advantages of the trial hole are that it is done at minimum cost, quickly, and the material can be seen in bulk. Gravel and sand can be given a quick test by driving a fork or bar into the surface; clay can be checked by a pocket penetrometer. Other tests will be described later in this chapter.

Figure 2.2 Pressure bulbs under single and multiple foundations.

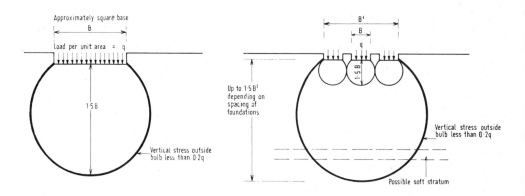

Auger borings

The auger is a corkscrew-like tool which can wind its way into the soil and either force it to the surface if power driven or take it out in small bites if hand operated. Hand augering can be extended in clay or a sandy clay or a clayey sand to about 6 m, but much greater depths can be achieved with power. Augering is not much use in gravels or cohesionless sands where the sides of the hole would collapse and is not effective where large cobbles or claystones might be encountered.

Figure 2.3 A typical shell boring rig at work on a housing site in Richmond, Surrey. Note the array of tools around the machine and the prepared undisturbed sampling tool in the background.

Figure 2.4 A close up of the assembly of the standard penetration test equipment in the rig of Figure 2.3. The heavy cylindrical weight is pulled up by the winch to a prescribed height and then released. The driving rod is marked by chalk at 75 mm intervals to enable the operative to register the number of blows for each 75 mm penetration.

Percussion borings

This is probably the most used technique as the boreholes can be lined. If necessary the boring depth can be extended to about 20 m with a casing and to great depths if an unlined bore such as in clay. Casings are usually required, and the sections are screw-coupled and driven by the excavating tools into the section of the bore just excavated. A variety of these tools are available both for boring and sampling. Clays are usually excavated by a clay cutter or 'shell', the former having a steel tube with a cutting edge, the latter an added flap valve near the cutting edge. This shell is also used for gravels and sands and the technique generally is sometimes known as 'shell boring'. Various cutters and chisels are available to deal with buried concrete or boulders. Water is often added to the casing when excavating in gravel but it should be used as little as possible and recorded on the borehole chart. Bore diameters are 150 – 200 mm. Rig dimensions about 6 × 2·5 × 5 m high. A typical machine is shown in Figures 2.3 and 2.4 and consists of a tripod and winch. The winch does all the work, labour required being the driller who controls the winch as it drives the cable up and down dealing either with the excavating tools, the casing or the testing devices, and the driller's mate who assembles the tools and empties the spoil from the tube or shell. This is a dangerous operation and the visiting engineer or architect would be well advised to stand clear until the cable has ceased movement.

Sampling and testing

As the bore proceeds it is usual to take bulk samples in polythene bags at 1 m or 1·5 m. When in gravel or sand, dynamic tests are employed: the Standard Penetration Test and the Cone Penetration Test. These tests are undertaken with standard steel hollow cylinders with shaped ends which are driven by the winch, the cable lifting a weight, 64 kg, to a point where it is trigger-released and falls through a standard dimension, 750 mm, down a rod connected to the cylinder. The number of blows is recorded for each 75 mm of penetration through a distance of 450 mm. In positions where the rig cannot be accommodated it is possible to carry out the test by hand (Figure 2.5).

In clay, undisturbed samples are taken at about 1·5 m intervals. The method consists of a sample tube with a cutter screwed to the lower end, an extension piece at the top and a release valve at the tip to allow air from the sample tube to escape as it is replaced by the clay sample (Figure 2.6). The assembly is withdrawn after the sample has been obtained and the end of the tube waxed and capped off. This enables the laboratory to have a number of samples in the condition they were in at the time they were taken, although the very action of driving the cutter down and forcing the clay into the sampler must have some effect on the sample.

Another test for clay is the vane test. This is more easily carried out close to the surface, as in a trial pit, mentioned earlier, but can also be applied in a borehole. The clay cannot be too hard as the vane with four blades has to be driven below the excavation surface and rotated, thus measuring the shear strength of the clay. The

Figure 2.5   Using the standard penetration test
equipment in chalk at a property
development site in Leatherhead, Surrey.

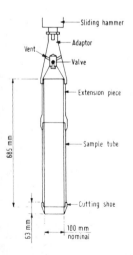

Figure 2.6   Typical undisturbed sampling tube for
cohesive soils.

rotating device at the head must incorporate a spring balance or alternative measuring device.

Other field tests

The static Dutch Cone Penetration Test is used widely in European countries in cohesionless soils. The penetration head is connected by steel tubes to a jacking device at the surface. First the head only is pushed down about 80 mm, then the sleeves are pushed down, engaging with the head and moving about 120 mm. This gives a resistance to the head in direct contact with the soil and a separate figure for the friction in the tube. The Dutch use this method almost exclusively.

The plate-bearing test is carried out at the bottom of a pit and consists of loading a steel plate and measuring the settlement in much the same way as in pile testing. The method is described in BS CP 2004.

The largest diameter plate possible should be used in order to try to take account of discontinuities in the soil. 300 mm diameter is a minimum. The maximum size will be dictated by the cost of applying the load, which should be about three times the theoretical load to satisfy the design. A 900 mm diameter plate design loaded to 125 tonnes per square metre would require a test load of about 80 tonnes. Although some indication will be given of the settlement under this loading, in the actual building the size of the footing loaded to a similar intensity will have a much larger stress bulb having an effect at greater depth. To obtain a meaningful answer it is necessary to multiply the settlement by the ratio of footing width to test plate width in clay, and in cohesionless soils it is necessary to extrapolate the results from a number of different plate sizes.

The water level has influence on the bearing value of sands and gravels, so that the plate test in which the pressure bulb does not reach the sands and gravels below water level may give an erroneous impression of the effect of the much larger building foundation, where the pressure bulb will extend below the water line.

Furthermore, the cost of handling Kentledge is high so that the test is costly. In the event that rock is known at, say, 10–12 m below the surface, it may be cheaper to bore down and put in rock anchors. The test is probably at its best on soils which are not readily assessable by other tests, for example, broken shales, chalk in which rocklike lumps are surrounded by chalk putty and other weathered soils which are almost impossible to sample.

Fill materials are also best suited to the plate test, but if there are large variations in the density of the fill no worthwhile result will be obtained.

There is no point in doing the test at all unless it is conducted in the manner of BS CP 2004 with a proving ring used to measure the load applied by the jack and a dial gauge with an accuracy of 0·05 mm to measure downward movement and upward recovery.

The Menard pressuremeter is used to determine the deformation characteristics of soils and rocks, its action being based on water pressure in rubber cells expanding

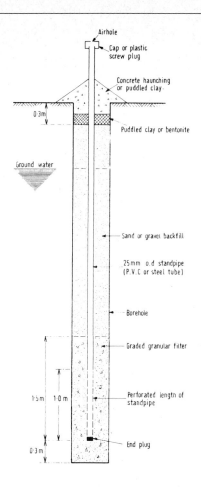

against the walls of the borehole. The geoprobe does much the same thing with gas pressure.

Pumping tests can be made when movement of ground water is important. The water table can be monitored by use of a perforated pipe sealed in the borehole (Figure 2.7).

Electrical, seismic or magnetic methods covered by the general description of 'geophysical' are used to determine the change in strata with markedly different properties. Great expertise is required and the methods are rarely used for surveys in connection with the building industry.

A typical borehole log is illustrated in Figure 2.8.

Figure 2.7 Standpipe sealed in bore hole to measure the movement of the groundwater over a period.

Figure 2.8 Typical bore hole log at a property development site in Whitton, Middlesex. Note the water level has risen to within 4·45 m of the surface after withdrawal of the casing.

CONTRACT & Location	HIGH STREET, WHITTON.						BOREHOLE No.: 1A		
CLIENT Cons. Engs.:	Bewsley Properties Ltd., Allan Hodgkinson and Associates.						REPORT No.: 1802/DW		
Method of Boring	Cable Percussion - 200 mm diameter - cased to 6.50 m.						Boring Started: 10.12.79 Boring Finished: 11.12.79		

GROUND WATER			Date	10.12.79	11.12.79	11.12.79	11.12.79
Water Strikes	Rate of Inflow:	Sealed off at:	Time	17.00	08.00	12.00	12.20
1st: 1.60	Fast	2.00	B/hole Depth	6.50	6.50	12.00	12.00
2nd: 8.30	Medium	Not sealed	Casing Depth	6.50	6.50	6.20	6.20
3rd:			Water Level	Dry	Dry	Dry	4.45

Remarks: Service pit dug at alternative borehole position for 1 hour. Water added to borehole from 3.1 m to 6.2 m to assist drilling. Chiselling on claystone at 8.3 m for 0.5 hour.

Samples		Depth (m)	S.P.T. (N)	Scale: 20mm=1m		O.D. =	Description
Ref. No.	Type			Legend	Depth		
9381	J	0.50					Made ground (soft dirty brown clay fill with stones, concrete fragments and brick fragments INFILLED TRIAL PIT.
9382	J	1.50					
9383	J	2.00			1.90		
9384	U	2.10 - 2.55					Firm brown and light grey sandy clay with scattered gravel.
9385	J	2.80			2.55		Medium-dense brown clayey sand with occasional gravel.
- 9386	CPT B	3.30 3.30	65		3.10		Dense gravel with some brown sand.
- 9387	CPT B	4.30 4.30	69				
- 9388	CPT B	5.30 5.30	34				
9389	J	6.25			6.20		Stiff brown clay.
9390	U	6.30 - 6.75			6.35		Stiff grey fissured clay. Claystone at 8.3 m.
9391	J	6.90					
9392	U	7.05 - 7.50					
9393	J	8.00					
9394	U	8.60 - 9.05					
9395	J	9.50					

Key: U = Undisturbed
B = Bulk
J = Jar
W = Water

Wembley Laboratories Limited

2.3   Laboratory Testing

As mentioned earlier, the site investigation contractor submits with his tender a list of the tests which his laboratory will carry out and the cost of each type of test. The usual tests employed by the soils engineer and laboratory technicians are as follows

1   Visual examination (all samples)
2   Natural moisture content (all undisturbed samples)
3   Liquid and plastic limits (clay)
4   Particle size distribution (sand and gravel)
5   Unconfined compression (clay)
6   Triaxial compression (clay and soft rocks)
7   Shear box (general)
8   Vane test (clay)
9   Consolidation (clay)
10   Permeability (sands and gravels and to some extent clays and silts)
11   Chemical analyses (general but mostly clay).

The visual examination is the confirmation or otherwise of the drilling foreman's record of his bulk samples. There is not much point in doing moisture content tests on disturbed samples as there will have been a loss of moisture from the original condition. In gravels there may actually be more moisture because of the use of water in the drilling operation.

Liquid and plastic limit tests were often omitted from investigations on simple structures but the N H B C requirement with respect to testing for clays with a high moisture movement makes this test essential where a registration with that organisation is required.

Particle size is determined by sieving with a set of standard size sieves. The grading curves resulting therefrom give little extra value in determining bearing pressure but are essential in the consideration of permeability. The shear box test (Figure 2.9) provides information on the shear strength of the soil and can be used to assess the ultimate bearing capacity.

Figure 2.9   Diagramatic arrangement of the shear box.

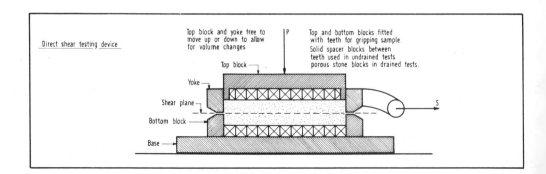

A simple quick test which can be applied to clays is the unconfined compression test and this can be carried out on site. It gives a value for apparent cohesion equal to half the shear strength. The main method of testing in the laboratory is the triaxial compression test (see Figure 2.10). Various lateral pressures are exerted on the sample from which it is possible to determine the value of c, cohesion and φ, angle of internal friction, represented by the diagram (Figure 2.11). This leads to the equation s = c + q tan φ which will be referred to later.

The test can be in three types: undrained, consolidated–undrained and drained. In the undrained test the specimen is not allowed to drain during the application of the three-dimensional pressure and the pore pressure is not allowed to dissipate during the test. This procedure reproduces the condition which occurs when the soil underneath the base is subjected to load or when earth is removed from an open or sheeted excavation.

Figure 2.10 One of a battery of three triaxial machines. The samples are tested simultaneously at three different external pressures.

Figure 2.11 The C/φ diagram. The Mohr circles are constructed from the results of the triaxial test.

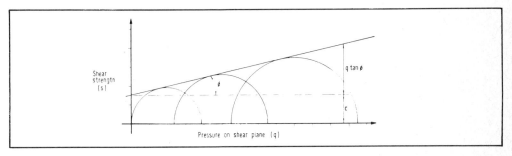

In the consolidated undrained test the specimen is allowed to drain while the overall pressure is applied, allowing the specimen to consolidate fully during the test.

The drained test allows drainage of the pore water during the test and corresponds to the conditions of long-term applications of the load.

The vane test has been described as a field test but can equally be carried out in the laboratory. Consolidation tests are used to calculate the rate and magnitude of consolidation below foundations. The sample is enclosed in a ring and the load is applied in only one direction (Figure 2.12). The results allow the coefficient of consolidation c_v to be obtained and some measure of the rate of settlement of the structure can then be calculated. A full cycle of loading and unloading is used to draw a pressure voids ratio curve from which mv (the coefficient of volume compressibility) is obtained. This enables the size of consolidation settlement to be assessed. The apparatus employed is called an oedometer.

This test takes up to two weeks to perform.

The permeability test can be carried out in the laboratory but is less satisfactory than the field test.

Chemical analyses are carried out primarily to determine the level of attack on buried members, mainly the steel and concrete of foundations.

Figure 2.12 The œdometer. Cohesive samples are being tested under long-term load to obtain information to be used in settlement calculations.

ph values are needed for all buried members and chloride values in order to measure attack on steel and for sulphate content in attack on the concrete. The ph value is the measure of alkaline or acid content present in the water but it cannot be used to measure the amount of acid or alkaline material present.

2.4   The Principles of Foundation Design

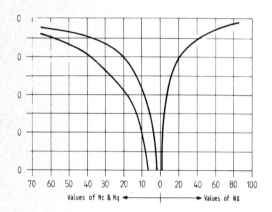

Figure 2.13   Terzaghi's bearing capacity factors.

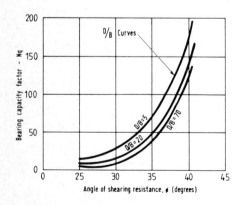

Figure 2.14   Berezantsev's bearing capacity factors Nq for end bearing of piles in cohesionless soils.

In Chapter 1 the difference between ultimate, presumed and allowable bearing pressures was briefly defined. The allowable bearing pressure is that which finally determines the foundation design, being some fraction of the ultimate bearing pressure. We now consider how the various properties defined from the site investigation and laboratory results are used in a theoretical process, usually referred to as soil mechanics, to assess the ultimate bearing pressure. When load is applied to a foundation, unless the soil is a solid unfissured rock, there is settlement which is part elastic, part plastic, depending on the nature of the soil. Assuming that a load of such magnitude is applied that the foundation continues settling, it will eventually run away without further load being applied, just as in the way a testing machine records the yield of a steel bar in a tensile test. At this point the ultimate bearing pressure Q_f has been reached. In cohesive soils this pressure is divided by a safety factor to give the allowable bearing pressure, though in cohesionless soils this pressure will be determined by the settlement at working load.

Various equations have been developed to add together the parameters which govern the ultimate failure. In the simplest form the equation reads

$$Q_f = N_c \, c f_c + f_q \, N_q \, p_o + \tfrac{1}{2} \, N_\gamma \gamma \, B_{f\gamma}.$$

where c = apparent cohesion

B = foundation width

γ = density of soil below the foundation

p_o = effective pressure of the overburden soil at foundation level

Nc, Nq and N_γ are bearing capacity factors depending on the value of φ.

f_c, f_q and f_γ are shape factors.

There are more complicated versions of the equation which bring into consideration depth, load inclination, inclination of underside of foundation and ground surface inclination.

For a strip foundation width B $f_c = 1·0$ $f_q = 1·0$ $f_\gamma = 1·0$

For a circle diameter B $f_c = 1·3$ $f_q = 1·2$ $f_\gamma = 0·6$

For a square side B $f_c = 1·3$ $f_q = 1·2$ $f_\gamma = 0·8$

For a rectangle sides short B long L $f_c = 1·0 + 0·3 \dfrac{B}{L}$ $f_q = 1 + 0·2 \dfrac{B}{L}$

$$f_\gamma = 1 - 0·3 \dfrac{B}{L}$$

The N_c, N_q and N_γ values are shown in Figure 2.13.

The N_q values are shown in Figure 2.14 for the base resistance to piles.

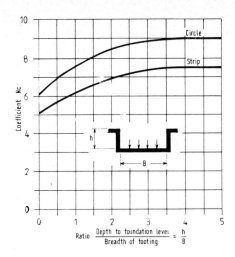

Figure 2.15 Skempton's bearing capacity factor Nc for deep foundations.

In deep foundations the skin friction on the sides of the foundation may become a considerable addition to bearing capacity and Figure 2.15 shows values proposed by Skempton.

Ultimate bearing capacities for strip foundations and the circle or square are given below:

$$Q_f = cN_c + p_o (N_q - 1) + \gamma \frac{B}{2} N_\gamma + p. \text{ (strip)}$$

$$\text{and } Q_f = 1 \cdot 3c \, N_c + p_o(N_q - 1) + 0 \cdot 4 \, B \, \gamma \, BN_\gamma + p \text{ (circle or square)}.$$

The value p is only relevant when the equivalent density of the displacing foundation is less than the density of the soil. Here the soil weight contributes to strength by holding down the soil which might be squeezed out from under the foundation. With a solid concrete base at density, say, 24 kN/m³ and a soil at, say, 19 kN/m³ it is usual to exclude the self weight of the base and ignore the p factor.

2.4.1   Cohesionless Soils

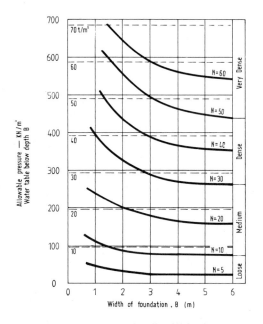

Figure 2.16 Chart for assessing allowable bearing pressure by the standard penetration test in cohesionless soils. N values are the number of blows per 300 mm of penetration.

With cohesion absent the N_c term becomes zero, leaving

$$Q_f = p_o (N_q - 1) + \gamma \frac{B}{2} N_\gamma \text{ strip}$$

$$Q_f = p_o (N_q - 1) + 0 \cdot 4B \, \gamma \, BN_\gamma \text{ circle and square.}$$

The standard penetration test, previously described, leads to the proposals by Terzaghi and Peck illustrated in Figure 2.16.

The chart should be interpreted as follows:

With dry or moist sand with the water table depth B below the foundation use the chart directly. The allowable pressure is equal approximately to $10 \times N$ kN/m².

For chalk (but subject to seeing the chalk in bulk) the allowable pressure is approximately equal to $20 \times N$. In the case of saturated sand, shock may cause liquefaction where N is less than 5. When N is greater than 5 the chart can be used but modified so that, if df is the depth from surface to underside of foundation, when df is small the pressure should be reduced by 50% and when $\dfrac{df}{B}$ is approaching unity the pressure should be reduced by 33%.

When N is less than 10 and df < B < 2m the pressure should be checked against shear failure.

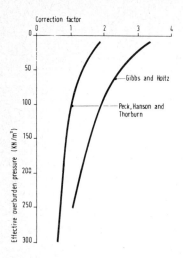

Figure 2.17　Correction factors for standard penetration test by Thorburn and Gibson and Holtz.

For fine or silty sand if N is greater than 15 use a modified value
$15 + \frac{1}{2}(N - 15)$.

The above rules are based on a total settlement of the foundation of 25 mm. There are two modifications to the chart figures which may be applied.

The first is an increase proposed by Gibbs and Holtz and by Thorburn who suggested that the penetration test seriously underestimates the relative densities of cohesionless soils at shallow depths. Their modification factors are shown in Figure 2.17. The more conservative Thorburn values are to be preferred.

The second variation relates to the presence of water below the base. The usual approach is to assume that the pressure derived from the N value be halved when the ground water level is close to the underside of the foundation.

This is not universally accepted as some engineers feel that the test results, as the tool drives into the wet soil, automatically allow for the bearing pressure reduction.

2.4.2　Cohesive Soils

Most clay soils are saturated and behave in a purely cohesive manner, i.e. $\phi = 0$ provided that no moisture is squeezed out under load. This is the usual assumption for building foundations where the load is applied comparatively quickly as opposed to, say, high dams.

Where the decrease in the moisture content does take place there is an increase in shear strength. This facility cannot be applied to a structure sensitive to settlement but could be made use of in the design of more flexible structures, e.g. steel storage tanks.

The earlier formulae simplify to $Q_f = cN_c + p$.

The standard penetration test may be applied in clay to give an approximate cohesion (Table 2.1).

Table 2.1　Relationship between the standard penetration test value N in a cohesive soil and allowable bearing pressure.

N	Very approximate apparent cohesion (kN/m²)	Usual description of clay
< 3	< 8	very soft
3–6	18–35	soft
6–12	35–70	firm
12–23	70–140	stiff
> 23	> 140	very stiff or hard

C − φ soils

This is the term used to describe soils which are intermediate between the cohesive and the cohesionless usually mixtures of gravel or sand with clay. It is possible to obtain samples and test as for clay and use the N_c, N_q, N_γ equation to determine the ultimate bearing pressure.

2.4.3 Factor of Safety

All the previous paragraphs have dealt with the ultimate bearing value and once this has been determined a judgement then has to be made on the application of a safety factor linked to the probability of settlement and differential settlement; this will be considered next. Values vary from 2 to 3 depending on the confidence derived from the direct examination of the soil, the test results and calculations and the sensitivity of the structure.

2.4.4 Ground Stress and Settlement

When a smooth rigid foundation loads the ground at its surface the pressure below the foundation varies as shown in Figure 2.18. As depth to underside of foundation increases the pressure below the bases in the sand, gravel and intermediate soils tend to a uniform pressure. But in clay a large foundation such as raft, or raft on piles, tries to settle to the centre of the foundation.

Simple approaches may be made in preliminary calculations. These assume that the load spreads through the soil at an inclination of 1 in 3 to the vertical (Figure 2.19).

Figure 2.19 Simple formula for calculating the width of load distribution at depth.

Figure 2.18 Distribution of pressure under an unyielding foundation on clay and on gravel.

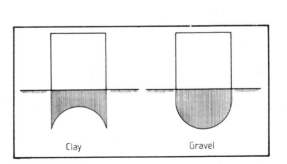

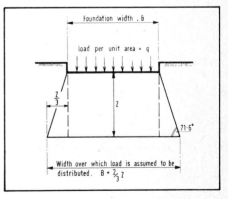

The theoretical approach is to treat the soil as a uniform elastic body employing the Boussinesq formula (Figure 2.20). With assumptions of load applied at the surface and the soil of infinite depth.

Figure 2.20 Boussinesq's equation from which the vertical stress in the soil can be calculated at various depths below the underside of a foundation.

$$\sigma_z = \frac{3Q}{2\pi z^2}\left[\frac{1}{1+\dfrac{r}{z}}\right]^{1/2}$$

Where Q = concentrated load

    z = vertical distance between point considered, N and underside of foundation

    r = horizontal distance from N to be line of action of load

    σ_z = vertical stress in the soil.

Variations to this theory allow for loads applied below the surface and for other assumptions about the soil. These theories represent the behaviour of clays reasonably well, but for sands and gravels the lateral distribution of concentrated loads is produced to a lesser degree. The stress distribution under a flexible circle and strip foundation are shown in Figure 2.21.

Obviously the foundation can only be treated in this way if the bulb of pressure from an adjacent foundation does not overlap.

Figure 2.21 Pressure bulbs under uniform circular and strip loads.

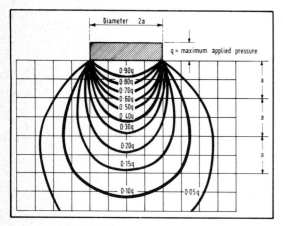
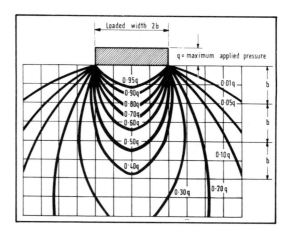

2.4.5 Estimation of Settlement

Assessment of the extent of settlement is a very complicated procedure requiring some experience in soil mechanics and is here considered only in principle. In any assessment it is necessary to decide the applied long-term load. This will vary according to the type and use of the building although in all cases it is obvious that the full self weight including the foundation weight must be employed. Superimposed load is a different proposition. At one end of the scale residential loading is nearer 0·5 kN/m² than the design load of 1·5. In office blocks usually designed for 5 kN/m² apart from file storage areas which might locally reach 10 kN/m², the average load in the remainder of the building is around 2·5. In multi-storey warehouses a different situation applies: the design load may be 15 kN/m² which will more likely be 20–25 kN/m² locally with the aisles left for movement nominally ·0 kN/m² but probably still averaging the design figure of 15 kN/m² for the whole floor area. Wind loading comes in gusts and in the time scale of the period over which the settlement takes place can be ignored.

The soil profile must be examined to the depth at which negligible stress is applied (Figure 2.22). It may well be that a compact gravel will present a presumed bearing capacity of 350 kN/m² in a strata depth of 1·2 m but below this may be a clay with not only a low presumed bearing capacity but also a high settlement characteristic. This clay may vary with depth so that in calculating the settlement it would be necessary to divide the clay in depth into layers with different settlement characteristics (Figure 2.23).

Figure 2.22 The soil profile must be examined to a depth at which the vertical stress is negligible.

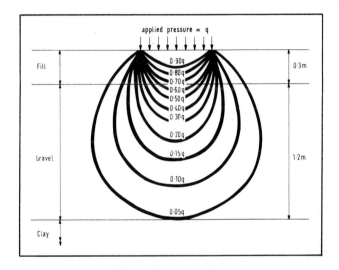

Figure 2.23 In variable soils the different layers with different cohesion values need to be considered in any consideration of settlement.

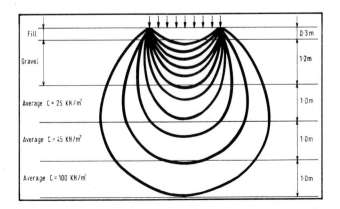

2.4.6   Cohesionless Soils

Two methods are available for estimating the settlement of a cohesionless soil which are based respectively on the standard penetration test and on the static cone penetration test. The former attributed to Menzenbach depends on a rather crude relationship between the penetration test results and the deformation modulus of the soil. In the estimate of settlement, the value of Poissons ratio must be assumed and the range of values suggested in turn suggest a wide scatter of settlement measurements. The static cone method of De Beer and Martens involves more complicated mathematics in order to derive the settlement measurement and is thought to give rather overestimated figures; the bearing pressure recommended therefrom can be increased for design purposes, by as much as 50%.

2.4.7   Cohesive Soils

For cohesive soils the immediate settlement is usually calculated using the previously described Boussinesq theory. Values must be assumed for the modulus of elasticity, E, and for Poissons ratio, usually taken as 0·5. Settlements are usually calculated on Terzaghi's theory of consolidation modified by Skempton and Bjerrum. Where there is a limited compressible stratum, Steinbrenner's method may be used, the same process being applied when there are a number of strata with different properties. For deeper foundations a reduced settlement occurs using factors of up to the 50% suggested by Fox. An approximation for a clay layer is the product of the thickness of the layer, the nett stress increase qm and the coefficient of volume decrease mv determined from the laboratory test. This settlement is determined by a factor which varies from about 0·2 to 1·2 depending on the type of clay and its geological history.

As calculations relate to an infinite time for the settlement, it is usual to take 90% of the calculated figure.

The time for settlement to occur may be calculated from the Cv coefficient provided that the length and nature of the drain path for the clay pore water under load is known. There are various computer programs to lessen the effort of the extensive stress and settlement calculations, though of course they in no way improve the accuracy of the result. The reader could be forgiven, having regard to all the assumptions which have to be made (often from borehole samples which may be representative of only a small area of the strata), for wondering why the experts do not guess the answer in the first place. In fact it is unlikely that the calculations will provide answers better than + or −50%, and all that is being calculated is the order of settlement and time for the settlement to occur. The results, if not accurate, are valuable in the consideration of relative settlement.

2.4.8 Other Load Influence and Area Affected

When making an assessment of settlement, the adjacent loading of the ground must not be overlooked. For example, the load on a foundation of a warehouse with a 30 m span and 6 m column module will be of the order of dead load 50 N and superimposed load of 67 kN, of which only, say, 20 kN need be considered for settlement, making a total load for settlement purposes of only 70 kN. This might be raised to 130 kN if brickwork and a ground beam are employed on the perimeter (Figure 2.24). Having designed the foundation on the full load of 177 kN, say 1 m sq, for an allowable pressure of 200 kN/m² the stress in the soil immediately below the foundation would be only 177 kN/m². The floor of the warehouse, however, could have been designed for 50 kN/m² and would apply over the whole bay related to this foundation, i.e. 15 m to midspan and the 6 m module. Obviously this heavily loaded area will have as much effect on total settlement as the foundation itself. It must also be appreciated when dealing with clay that the settlement is effective well outside the confines of the building.

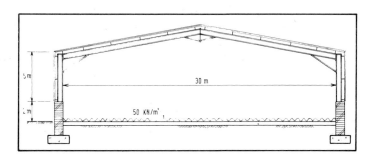

Figure 2.24 Typical warehouse section showing how a heavy floor load can outweigh the settlement condition of the individual footing.

2.4.9 The Capacity of the Building to Accept Settlement

Obviously in any consideration of soil settlement the type of structure being considered must always be borne in mind when determining the final allowable bearing pressure to be used in the foundation design. The problem in the warehouse described in the previous paragraph is totally different from the twenty-storey office building and again totally different from the four-storey load-bearing masonry structure of a block of flats. The framed structure requires different assessment from the load-bearing masonry structure as the flexibility of the former allows greater differential movement.

Assessments have been made of the tolerance of different types of structure to differential movement (Figure 2.25). Some advice is given on distortion and settlement in a paper by Skempton and MacDonald. They refer to damage limits for load-bearing walls and panels in traditional framed buildings (Table 2.2). For design purposes they suggest that a factor of safety of 1·25–1·50 be

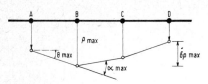

(a)   Definitions of settlement ρ , relative settlement $\delta\rho$, rotation θ and angular strain $\propto$

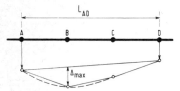

(b)   Definitions of relative deflection Δ and deflection ratio $\Delta/_L$

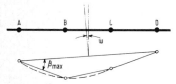

(c)   Definitions of tilt ω and relative rotation (angular distortion) β

Figure 2.25 **Definitions of foundation movement, settlement, relative deflection, tilt and relative rotation.**

Table 2.2 **Criteria for angular distortion and settlement without damage to the structure.**

applied which would suggest a limit of angular distortion of 1:450. The bare frame without considering the infill could be double this amount at 1:225. Other investigators have arrived at figures of 1:750 and 1:150 indicating the lack of precision possible and therefore the desirability of erring on the safe side. Considering the partitions inside the building figures of 1:300–1:500 are recommended.

For unreinforced load-bearing walls Burland and Roth recommend a deflection ratio Δ/L at length divided by height $\dfrac{L}{H}$ equals unity of $0{\cdot}4 \times 10^{-3}$, at

$\dfrac{L}{H}$ equals five of $0{\cdot}8 \times 10^{-3}$ where cracking is caused by the wall sagging and

$0{\cdot}2 \times 10^{-3}$ and $0{\cdot}4 \times 10^{-3}$ respectively where the cracking is caused by hogging.

Criterion		Isolated foundations	Rafts
Angular distortion		$\dfrac{1}{300}$	$\dfrac{1}{300}$
Greatest differential settlement:	clay	45 mm	45 mm
	sand	32 mm	32 mm
Maximum settlement:	clay	76 mm	76–127 mm
	sand	50 mm	50–76 mm

General References

Skempton, A.W., *The Bearing Capacity of Clays*, Building Research Congress Division, 1, 1951.

Fox, E.N. 'The mean elastic settlement of a uniformly loaded area at a depth below the ground surface', 2nd Int. Conf. on Soil Mech., vol. 1, 1948.

Skempton, A.W. and MacDonald, 'The allowable settlement of buildings', *Proceedings of ICE*, Part 3.5, 1956.

Terzaghi, K., and Peck, R.B., *Soil Mechanics in Engineering Practice*, Wiley, 1968.

Burland, J.B. and Wroth, C.P., *Settlement of Buildings and Associated Damage* (Conference on Settlement of Structures), Pentech Press, 1974.

Tomlinson, M.J. *Foundation Design and Construction*, 4th edn, Pitman, 1980.

BS CP 5930: 1981 Code of Practice for site investigations, BSI.

Skempton, A.W. and Bjerrun, 'Settlement analysis of foundations on clay', *Geotechnique* 7, to be published 1987.

Chapter 3

Foundation Types

The criteria for a successful foundation are that it should be carried to a minimum depth at minimum width without exceeding the allowable bearing pressure of any soil layer beneath it; have settlement consistent with the supported structure; be at depth below the ground level to avoid natural ground movements due to frost, heat or moisture; regard be given to the ability of the contractor to construct it with a minimum of complication. At first sight a formidable task. However, with a few years of experience, about 95% of foundation situations lead to a ready solution.

3.1 Spread Foundations

3.1.1 Strip Footings

Examination of buildings where underpinning has become necessary has revealed some intriguing foundations of the late years of the nineteenth century. Some of these have survived eighty years or more, and include several long hot summers (Figure 3.1).

The modern strip footing (Figure 3.2) is either a thin strip of concrete placed in the base of a trench excavation from which the brickwork commences or a narrower trench completely filled with concrete. Ideally it should be kept above the water line; a wider trench built in the dry is preferable to a narrow one involving pumping.

Figure 3.1 Old types of foundations still giving service today.

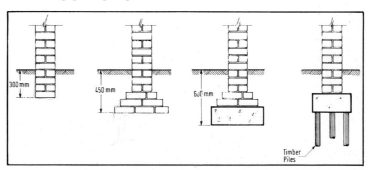

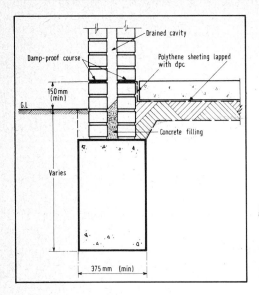

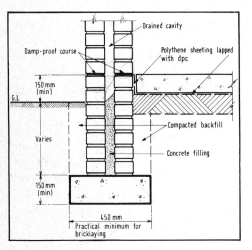

Figure 3.2 Alternative methods of constructing strip foundations.

Figure 3.3 Foundation arrangement for three- to four-storey flats on sloping chalk in Reedham, Surrey.

The economics are determined by a number of factors. In a trench deeper than 900 mm, the depth usually demanded in non-bloating clay areas where trees are not a problem, a width of nearly 600 mm is required to build a typical 260 mm cavity wall irrespective of the ground pressure. If the ground pressure is such that a 400 mm wide trench can be dug and filled with concrete the same day this form of construction wins. Here we are considering a situation typical to a house or to flats. The ground conditions may be such that a wider strip is required to meet the lesser quality soil at a lower level and in this case the reinforced wider strip will be preferable to the trench fill (Figure 3.2). Soil layers of different types at lower depths may well influence the form of construction. Figure 3.3 shows the section of the ground in a large residential development at Reedham in Surrey in this case of a three four-storey block of flats. The clayey flint deposits on the chalk could have carried 200 kN/m² but avoidance of differential settlement between the two ends of the building demanded that the entire foundation should go to the chalk. Here the allowable capacity was 320 kN/m² and, using an excavator with a long reach, trench/fill foundations only 450 mm wide were able to be taken to a depth of 6 m. An alternative solution could have been to found the four-storey section on the clayey flint and the three-storey section on the chalk had it been possible to make a movement joint through the full height of the building.

Strip footings are usually employed in load-bearing masonry construction where there is a fairly consistent load applied to the ground in all main walls of the building. Regulations require a minimum strip thickness of 150 mm with a projection from the face of the masonry equal to the thickness of the footings. The footing can fail either in bending or in shear (Figure 3.4). A shear failure is more likely where the footing has been widened and reinforced (Figure 3.5). Generally reinforcement is avoided by proportioning the thickness of the footing so that a 45⁰ line from the base of the masonry lies just inside the extremity of the footing (Figure 3.6). Increasing the width of a brick base

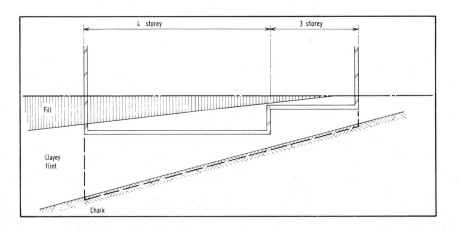

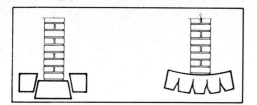

Figure 3.4 Foundation failure in shear and bending.

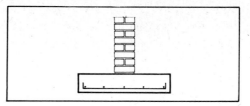

Figure 3.5 Reinforcement used to overcome the failures of Figure 3.4.

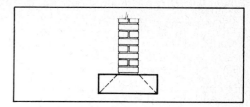

Figure 3.6 Foundation depth sized to allow load spread to lie within the foundation width.

follows the normal law of corbelling in that the 225 brick is stepped 225/4 for each course.

If the ground slopes then the footing can be stepped, in which case there are two rules to follow (see Figure 3.7).

While reinforcement in a strip foundation is a construction nuisance, quite apart from being costly, it may be needed to bridge over changeable spots in the soil, soft or hard.

Where a calculated foundation is needed then the design procedure is, as in Figure 3.8, the main transverse reinforcement being designed and the longitudinal reinforcement varying from about 10% of the main reinforcement 20% depending on the consistency of the soil in the length of the footing or whether some point load such as a chimney-breast or a ground beam at structure floor level needs to be spread.

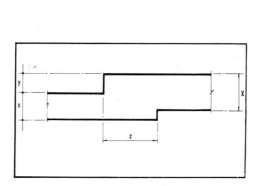

Figure 3.7 Stepping of foundation in accordance with the Building Regulations.

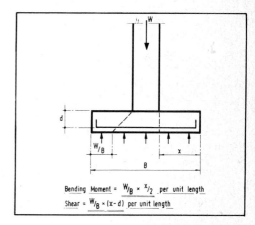

Bending Moment $= \frac{W}{B} \times \frac{x}{2}$ per unit length

Shear $= \frac{W}{B} \times (x - d)$ per unit length

Figure 3.8 Illustration of bending moment and shear calculation in a strip footing.

3.1.2 Pad Footings

There are occasions when the above-mentioned structures are founded on soil consisting of soft fill over a gravel worth say 320 kN/m². Here it would be uneconomical to excavate a deep trench in the fill, probably strutted, and the

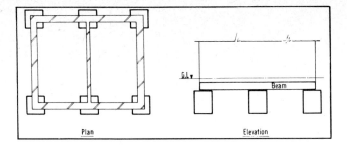

Plan   Elevation

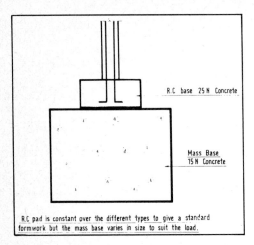

R.C base 25 N Concrete

Mass Base
15 N Concrete

R.C pad is constant over the different types to give a standard
formwork but the mass base varies in size to suit the load.

Figure 3.10   Standard RC pad used with varying mass base
             sizes to accommodate varying column loads in
             a structure.

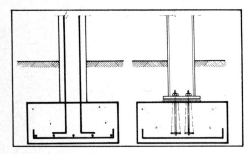

Figure 3.11   Typical arrangements of foundations beneath
             a reinforced concrete and steel stanchion.

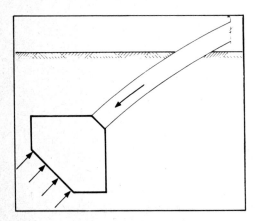

Figure 3.12   Foundation shaped to receive the resolved
             thrust from the structure.

answer is to use a ground beam under a length of masonry wall spanning to a pad foundation at a junction of walls or, if the wall is long, halfway along its length. The pad foundation can be more easily planked and strutted and concreted up to the underside of the ground beam (Figure 3.9).

Pad foundations are normally used for a structure with a frame in which there are clearly identifiable positions of structure load at ground level. Having regard to the concentration of load, the foundation may consist of high quality concrete (250 Kg per cu.m of concrete) if it is small in area or a high quality reinforced concrete pad on a lower quality mass concrete base if larger (Figure 3.10).

The column load may derive from a steel structure or a reinforced concrete structure resulting in a somewhat different detail in the first 600 mm depth of the base (Figure 3.11). Most columns, whether steel or reinforced concrete, are designed as simply supported primarily for ease of detailing but often because the soil is such that a very small rotation of the base would destroy the fixity. In the same way portal frames in either steel or reinforced concrete are designed with simply supported bases but because of the permanent outward thrust at the base of the column are provided with a tie across the width of the building. This not only avoids the soil behind the base having to provide passive resistance which may be negated by weather conditions, but also provides a safe-

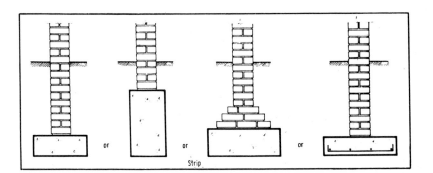

or   or   or

Strip

Figure 3.13   Alternative arrangements of strip foundations.

Figure 3.14   Alternative arrangements of mass foundations.

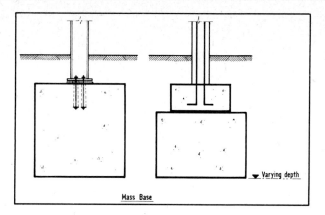

Mass Base

guard against future excavations, such as for drainage, immediately behind the base.

Where the base has to resist both vertical load and horizontal thrust, as in some bridge foundations, it can be shaped to receive the dead load resulting from these forces (Figure 3.12).

Design of strip and pad bases varies considerably (Figures 3.13, 3.14).

3.1.3   Eccentrically Loaded Pad Footings

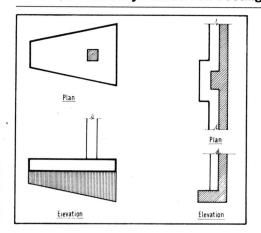

Plan

Plan

Elevation   Elevation

Figure 3.15   Trapezoidal and eccentrically loaded foundations.

There could be a situation in which the load on a pad has to be placed off centre. This could occur where a column load is combined with a retaining wall or where an adjacent structure prevents the base being placed symmetrically (Figure 3.15).

3.1.4   Continuous Pad and Beam Foundations

The continuous foundation is employed when a number of columns in line would, were their load spread longitudinally, have almost overlapping bearing areas. This type of foundation is usually overdesigned because of the number of variables involved, varying column loads, varying soil qualities and the

Figure 3.16 A rigid foundation beam under a number of column loads.

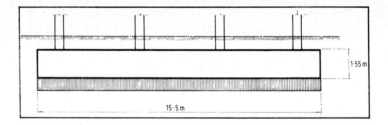

Figure 3.17 A flexible foundation beam under a number of column loads.

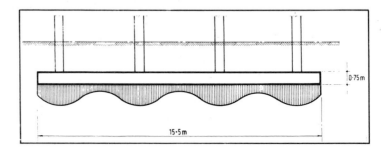

chosen depth of the foundation which may mean it is rigid or flexible. A rigid foundation design is shown in Figure 3.16. Computer fans will probably claim that they can provide an equally good design for a flexible foundation (Figure 3.17) by equalising the applied and ground reactions, but the result will be no better than the assumptions made about the soil properties.

The beam foundation is used in several situations. Figure 3.18 shows one column close to an adjacent property linked to the next column in line. Figure 3.19 shows two columns close together which are balanced on one foundation. Figure 3.20 shows a cantilever beam in an example where the new building

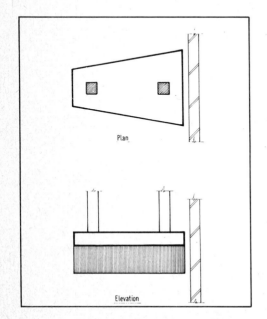

Figure 3.18 Column adjacent to existing property.

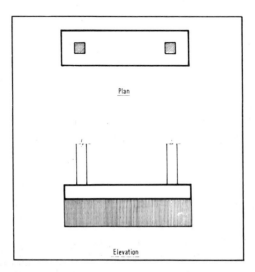

Figure 3.19 Two columns balanced on one foundation.

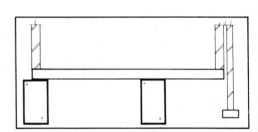

Figure 3.20 A cantilevered foundation.

foundation cannot be placed against the existing building. In this case it is essential that the holding down load, when the superimposed load acts in the worst possible position, still gives a factor of safety of 1·25.

3.1.5 Raft Foundations

Raft foundations under normal circumstances are used when pad foundations would occupy more than half the area of the building perimeter, though Tomlinson has pointed out that this is not necessarily good engineering as the quantity of steel needed to avoid excessive deflection of the raft, especially with unequal column loads, could be large and a more economical solution could be to excavate the site as though for a raft, construct pad bases and then back-fill between them.

A variety of structures described as rafts are shown in Figure 3.21, though in

Figure 3.21 Alternative raft arrangements.

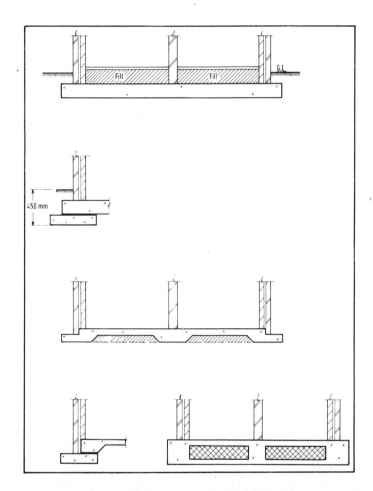

the opinion of the author several of them are nothing more than a series of strip foundations tied together with a ground slab. These are likely to be employed only in two-storey structures of the residential type with light loads, and simple strip foundation calculation can be employed.

Upstand or downstand beams forming a grillage with a substantial slab between them, or a really thick slab, stand some reasonable chance of minimising differential settlements. In these types of raft, design is difficult because of the complications in assessing the distribution of ground pressure and hence the bending moments and shear stresses in the raft. Designs can vary from the uneconomical to the potentially unsafe.

A pressure distribution which corresponds with loads as well as with the relative stiffnesses of soil and structure must be found. If the raft is flexible the loads will pass almost directly to the soil. As the soil becomes stiffer, loads are distributed to a greater extent between the points of application, but the overall deformation of the soil begins to impose a further redistribution. In the extreme case of a very rigid raft on clay, there will be a large concentration of pressure near the edges, as noted in Chapter 2, and hence a large overall bending moment which may even override local spreading moments.

Approach to raft design is very diverse. At one time it was assumed that ground pressure was distributed uniformly or varied linearly according to the eccentricity of the total load relative to the raft centroid. This can give too big a spread and yet not allow for the overall moment mentioned above; also, the moments and shears thus calculated are often unrealistic.

Professor A.L.L. Baker's soil line method was an early attempt to marry the structure and soil interaction, and large buildings in South Africa were designed successfully on that basis.

Other manual methods include the assumption of uniform or varying soil springs. Terzaghi's modulus of subgrade reaction is such a measurement of the soil behaviour.

In designing a project, there is a limit to the amount of calculation relative to time and therefore a manual computation can only go so far. The computer can carry out the calculation at enormous speed and, by an iteration process, reach a logical distribution of shears and bending moments in which each small section of the raft can be compatible with its neighbour. The final solution is, of course, no better than the parameters fed into the program so that the end result will still need careful and skilful interpretation.

Settlement rather than safe bearing capacity is the normal criterion of raft design, but preliminary sizing is done on a conservative bearing pressure, with a safety factor of 3, to allow for the occurrence of high pressures. The final pressure at the edges, where shear failure could occur, should be checked against this. Having regard to the area of the raft, the pressure bulb is large and total settlements can be high even though differential settlements may be reduced. Rafts are often used in conjunction with piles in high buildings, primarily with the object of minimising differential settlement, but some buildings have been designed in which the raft takes a portion of the load.

3.2 Piles

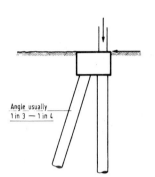

Figure 3.22 Illustration of a raking pile.

Piles are employed usually where, because of soil conditions, economic or constructional considerations, it is desirable to transmit loads to strata beyond the practicable reach of spread foundations. There may well be a high water table at a depth where spread foundations could be employed but where piling may provide a cheaper answer than the lowering of the water level. Piled foundations may also provide a satisfactory solution where a site is very restricted and large spread foundations could produce constructional complications.

Piles can also act in tension to resist uplift which might be caused by wind, or where a buoyancy condition by varying water levels might be created. It is possible to design piles to resist a modest horizontal force at their head but as the force increases raking piles will be probably be required (Figure 3.22).

Having regard to the possibility of differential settlement, piles and spread foundations are rarely used in support of the same structures.

3.2.1 Pile Types

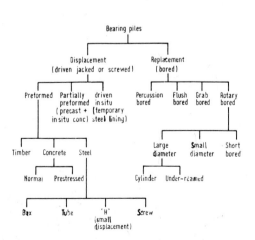

Figure 3.23 Analysis of basic pile types.

Figure 3.23 has been published many times and is a good explanatory picture of the different kinds of piles. Displacement means that a driving or pressure or vibratory technique is used to form the pile in the ground without recourse to excavation. Replacement means that soil is actually removed and either the resulting hole filled with concrete or, sometimes, a precast concrete pile is dropped in the hole and grouted in. One type of replacement pile is formed by vibrating stone aggregate into the hole and grouting this stone matrix.

Piles may be end bearing, rely on friction or adhesion developed along the sides, or may rely on a combination of both. In the last case it is usual to specify a factor of safety of 2·5 or 3·0 depending on whether load tests are carried out, but settlement of the pile must take place before the base load is mobilised. Large-diameter piles are thought of as being in excess of 600 mm diameter. Up to this diameter they can be constructed either by the 'baling' process or by machine. Beyond the 600 mm diameter, machines will be employed and, in clay, can be under-reamed subject to shaft diameter to as much as 6 m diameter. Piling is very much a specialised technique and many machines and pile types have been developed for both general and specific purposes. Figure 3.24 indicates the range of piles available.

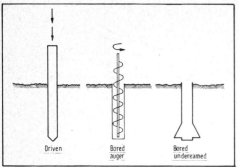

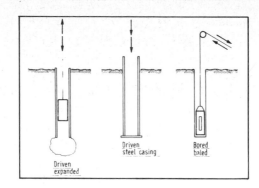

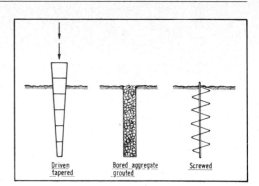

Figure 3.24 Alternative types of pile.

3.2.2 Pile Design

Displacement piles

The driven pile is hammered into the ground by dropping a weight on to its head through a protective assembly of a dolly and packing contained by a helmet. Alternatively, a steam or air-supplied piston may be used instead of the weight, The pile is driven to a 'set', i.e., a set of blows to 25 mm penetration. Many formulae have been put forward to calculate the set, the most widely used one being the Hiley formulae or derivatives of it.

$$R = \frac{Wh\eta}{S + \dfrac{c}{2}}$$

W is the weight of the hammer,
h is the distance through which the hammer falls,
η is the efficiency of the blow,
S is the penetration of the pile under each blow
c is the temporary elastic compression.

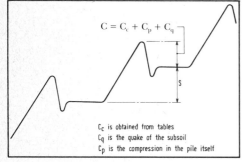

Figure 3.25 Diagram of the vertical movement of a pile shaft during driving.

When the hammer hits the head of the pile there is a compression of the dolly and packing and a reduction in the length of the pile. After impact, the pile recovers and the effects are expressed in the formula by coefficients. Measurements can be taken during driving by having a horizontal rail supported well outside the driving influence. By holding a pencil against the pile the compression and recovery of the pile can be registered (Figure 3.25).

The dynamic formulae are based on Newtonian energy theory which seems plausible. However, none of the formulae takes account of the soil properties or the long-term effects.

Suprisingly, a concrete pile will be damaged in tension rather than in compression as the stress wave travels back up the pile and this was one of the reasons for the introduction of the prestressed pile. The head and toe can be reinforced by links or spirals to accept the severe impact at these points.

Obviously there is no such problem with steel piles. Displacement piles are normally driven at centres no less than twice pile diameter to facilitate the driving of the pile.

Replacement piles

As described previously the replacement pile can be end bearing, rely on friction or adhesion along the sides, or a combination of both.

Design loads are determined by use of the standard N_c, N_q and N_γ coefficients defined in Chapter 2. Specialist piling contractors can usually undertake piling design, but it is generally preferable to use an independent engineer who can advise on the suitability of various types of pile. Contractors prepare schemes that suit their own methods, often making it difficult to assess the relative merits of different proposals, whereas engineers will try to produce one basic design and elicit truly comparable and competitive tenders from different specialists. Where this is not practicable alternative schemes may be prepared for which quotations can be obtained.

When a contract is placed, the engineer usually agrees final details with the specialist contractor in the light both of the conditions encountered during the work and of the contractor's method of installation; this avoids diminishing the contractor's responsibility. At this stage advantage can be taken of any features of the chosen system which permit economies on basic tender design.

Piling for foundations can be executed as part of the main contractor's work, under a separate contract before the main contract begins, or carried out as a nominated sub-contract. The main information to be given when inviting tenders is listed later in a typical condition of contract document. This includes a selection of general technical specification clauses, proposed by the Federation of Piling Specialists to produce comparable tenders for proprietary types of cast-in-place driven or bored piling. These clauses deliberately omit matters which should be a condition of contract, but are very informative on constructional aspects of this important type of piling.

Bored piles in which the shaft takes the majority of the load are placed at centres not less than three times the pile diameter. At two diameter centres the effective load is reduced by 40%.

3.2.3   Pile Testing

Pile tests are very costly and consideration should be given to the use of a higher factor of safety, and the resulting pile cost being compared with the cost of the testing. This is not always relevant with the present system of Local Authority involvement where engineering judgement takes second place to routine.

Testing may be static, with load applied through a staging of kentledge or jacking against tension piles or dynamic, in which there is a constant rate of penetration.

If an adequate number of tests to failure are made, say 1 in 50, the safety factor could be reduced to 2. An alternative testing method is to test a number of the piles to be used in the works to 1·5 times the design load. A settlement of 6 mm under working load is normally considered acceptable.

3.2.4   Settlement

Settlement of a single pile may be assessed from the pile test or by calculation taking account of the soil under the base having allowed for the skin friction up to the ultimate at working load.

When a group of piles is employed the settlement may be considered as an equivalent raft at two thirds of the pile depth for friction piles and at the bottom for end bearing piles.

Another consideration when employing groups of piles is that the entire group may fail (Figure 3.26).

Figure 3.26   Block failure in a pile group.

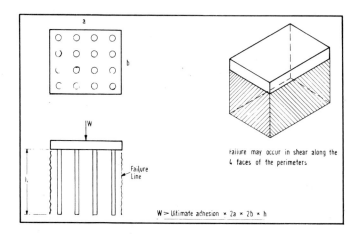

failure may occur in shear along the 4 faces of the perimeters

$W >$ Ultimate adhesion × 2a × 2b × h

3.2.5   Details – Replacement Types

Small bored piles. 300–600 mm diameter

Small bored piles are usually constructed by a tripod rig though one company uses a conventional crawler rig and vehicle mounted augers have been used for small house supporting piles. The process is very similar to the site investigation boring described in Chapter 2. The equipment consists of a tripod, a winch and a cable operating a variety of tools. The basic tools are a cylindrical shell with a cutting edge and a flap valve at the bottom usually employed in cohesionless soils, adding water down the bore, and a cruciform section tool with a small cylindrical cutting edge at the bottom usually employed in cohesive soils (Figures 3.27a, b). Other tools are used for breaking up clay stones or other hard objects in the ground.

In cohesionless soils the borehole must be stabilised by a steel casing which is withdrawn when the concrete is placed. It is normal practice to have a reinforcing cage the full depth of the bore in cohesionless soils and to a depth of 1·0 m below the clay level in cohesive soils (Figures 3.28 and 3.29).

Figure 3.27   Cutting tools for use in shell piling. (a) Type used in soft or cohesionless soils. (b) Type used in hard clays.

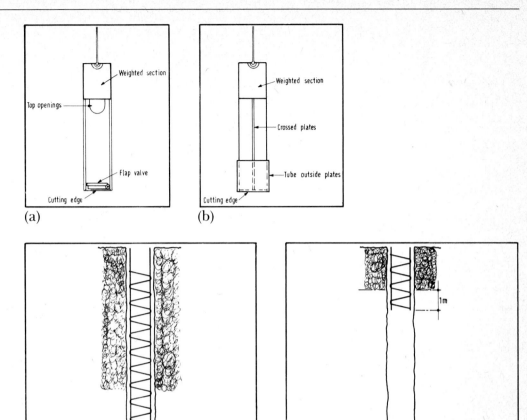

(a)   (b)

Figure 3.28   Reinforcing cage in a pile in cohesionless soil.

Figure 3.29   Reinforcing cage in a pile in a cohesive soil.

Figure 3.30   Tremie used for placing concrete under water. A sufficient head of concrete must be kept in the tube to ensure a continuous flow of concrete to the pile with the tremie bottom below the rising concrete level.

Concrete of a slump of 125–150 mm is employed and as it is difficult to judge the right amount of concrete to complete the pile to a level, having regard to the operation of withdrawing the casing, the concrete is usually specified to come up to ground level. In fact it does so and spills all around the pile head. The tripod pile has the least equipment on-cost and is very useful on small sites where it is more flexible than the larger machine-constructed piles.

In all bored piles if water cannot be cut off by the casing the concrete must be placed by 'tremie' (Figure 3.30). In this event the concrete slump should be increased to 175 mm and concrete mix to 1:1·5:3. The normal maximum depth to go with a casing is 20 m but it is possible to go deeper with an unlined bore.

Rotary bored piles
The rotary bored pile can vary in diameter from 300 mm to 3 m with under-reams up to 7 m. It can be crane mounted or lorry mounted and the larger types would obviously only be employed where a large number of piles are to be constructed.

Figure 3.31   Digging auger.

Figure 3.32   Bucket digger.

The potential of these piles can be seen when an under-ream of, say 6 m diameter in a hard clay at a working value of 400 kN/m^2 will carry over 1,000 tonnes and twice as much on rock chalk.

However, under-reaming is a slow process requiring a stop in the augering for change of tool and a slow process in the actual under-reaming operation. In clay it is often preferable to use a deeper straight-sided shaft. Under-reams are examined to see that no debris is present and this involves a protective shaft and a cage lowered to the base.

The boring tools are either of the auger or the bucket type. In the former the excavated material is disposed above ground by spinning the auger in reverse, in the latter the bucket disposes of the soils through its bottom door (Figures 3.31 and 3.32).

The boring device is attached to a shaft known as a Kelly bar, i.e. a square telescopic member which can reach depths of 60 m or more (Figure 3.33) driven by a horizontal spinner. The bar itself hangs from the mast of the crane or lorry.

One of the problems of the larger pile is the insertion of the casing. Vibration appears to be the better solution but the casing can be driven or twisted (by the Kelly bar). Alternatively a technique known as 'mudding in' is employed where a bentonite slurry is formed with the short auger both twisting and being leaned

Figure 3.33   Large bored under-reamed piling equipment at the British Library Site, St Pancras.   (a) Seen in its extended position, this under-reaming tool was developed by Pigott for forming 'bells' up to a maximum of 4·2 m diameter.   (b) Man-riding cage used for inspecting each pile shaft and under-ream.   (c) Reinforcement cage being positioned prior to lowering into a pile under-ream and shaft.   (d) Newer of the two Hughes KCA boring rigs on the site, seen here fitted with the under-reaming tool.

(a)

(b)

(c)

(d)

on by the rig. The process continues dragging down the casing. A further use of bentonite is to fill the hole with this thixotropic slurry, the casing then being inserted quite simply.

The same procedure of reinforcement cage is used providing about 0·5% of pile area in the vertical bars. The links are often welded to the main bars to make the cage manageable and, of course, the size of the cage means that a separate crane has to be employed. A concrete mix of 1:2:4 is usually employed with a 125 slump as the stresses in the shaft are mostly less than 4 N/mm².

Grout intruded piles

This type of pile employs a flight auger which bores to the required depth leaving the soil on the auger. Grout can then be forced down the hollow shaft of the auger and continues building up from the bottom as the auger with its load of spoil is withdrawn. Reinforcement can be lowered into the grout before it sets (Figure 3.34).

An alternative system used in granular soils actually uses the soil, mixing it with the pressured grout as the auger withdraws. Of course no reinforcement can be used in this type of pile.

Partially preformed replacement piles

This pile is particularly suitable for waterlogged ground or where there is

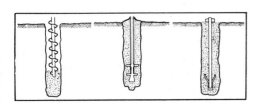

Figure 3.34 Grout intruded piles. (a) The grout is injected through the hollow shaft of the auger and displaces the spoil.   (b) The grout is injected as (a) but is mixed with the spoil by reverse action of the auger.   (c) In either (a) or (b) a threaded rod can be installed and tightened down.

movement of water in an upper layer of the soil. A hole is bored in the normal way and then precast annular sections are lowered into the hole connecting at the casing head as the preceding section reaches that position. With the full cored column in position, reinforcement can be placed and grout is then forced to the base of the column, displacing water and filling both the gap outside the column and the core inside the column (Figure 3.35). At a factory extension in Pontypool a hard marl was available at some 9 m depth but a soil layer above contained moving water. Rather than risk the cement being leached from the pile concrete, a precast pile was dropped into the hole in which grout had already been placed. A driven pile could have provided an adequate solution but the vibration would have affected the mechanical plant of the existing factory (Figure 3.36).

Figure 3.35 Stages in the process of constructing from precast concrete shells, a grout intruded reinforced pile.

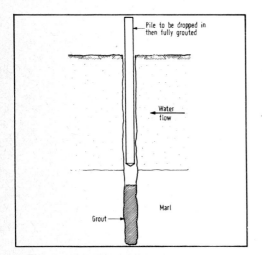

Figure 3.36 British Nylon Spinners Factory, Pontypool. Pile embedded in a prebored hole in the marl to avoid leaching of an *in situ* pile of concrete by water flow in the upper layers.

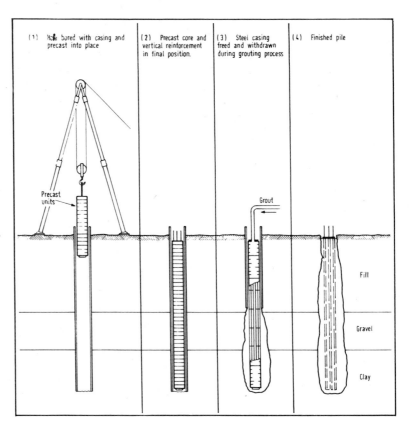

3.2.6   Pile Details – Displacement Types

The earliest driven piles were of timber but, except in remote parts of the world, they are now of reinforced concrete, prestressed concrete, steel and a variety of casings (steel or concrete, temporary or permanent) filled with *in situ* concrete. All except the temporary casing type lend themselves to piling through water.

Solid concrete piles

This type of pile may be either of reinforced concrete or prestressed concrete cast on site or in a factory. If the latter, its length would be limited to about 20 m by transport restrictions. The main disadvantages of this type of pile are that, with hard driving, damage may occur without detection and unless the founding soil top level is fairly uniform there can be a wastage of pile length, or worse still, the necessity to extend the pile in position.

Hollow concrete piles perform in a similar manner and for the same weight provide a greater resistance to lateral bending.

Solid concrete segmental piles

In order to deal with the extending problem the solid segmental pile was developed. This type has factory-produced segments with different proprietary shapes and methods of joining the segments on site; basically these consist of a male end to which the pile reinforcement is coupled and a female end to which the reinforcement of the next segment is attached. This type performs precisely as the solid concrete pile.

Hollow concrete segmental piles

One objection to the solid pile is the possible undetected damage. The hollow concrete segmental pile can be driven to a set and then filled with reinforcement and concrete. Not only does this allow hair cracks in the shell to occur without affecting the *in situ* core, but if the ground conditions are such that a higher strength and sulphate resisting concrete is called for, using this type of pile only the shell needs to have this specification (Figure 3.37).

Driven cast-in-place piles

There are several versions of the driven cast-in-place pile (Figures 3.38–3.40).

(a) Using a temporary steel driving tube. In this type of pile a steel tube is driven into the soil by a variety of rig types from the straightforward converted excavator with leaders to specially designed frames which can roll or walk. These latter have a considerable on-cost in being brought to the site. Drop hammers or vibrating hammers can be used to drive the tube to the appropriate depth usually determined by some variation of the Hiley formula.

(b) Figure 3.41. Here a lump of dry concrete materials is placed at the bottom of the tube and a cylindrical hammer is dropped successively causing the tube to be carried down with the concrete plug. The pile can

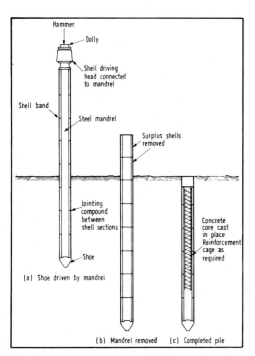

Figure 3.37   Precast unit shells driven to a set then filled with reinforcement and concrete.

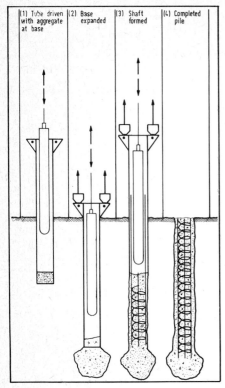

Figure 3.38   Driven cast-in-place piling in which an expanded concrete base is constructed.

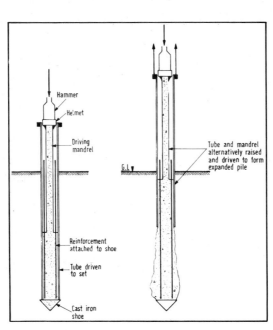

Figure 3.39   Stages in the formation of an alpha pile.

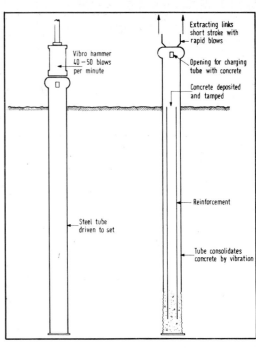

Figure 3.40   Stages in the formation of a vibro pile.

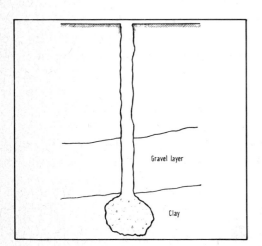

Figure 3.41   Driven cast-in-place piling where an expanded gravel is used to effectively continue the gravel stratum into the clay.

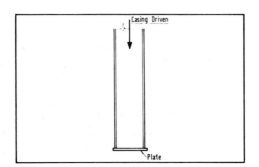

Figure 3.42   Driven cast-in-place piling driving a flat plate or shaped penetrating plate to a set.

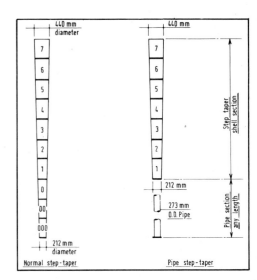

Figure 3.43   Steel taper piles.

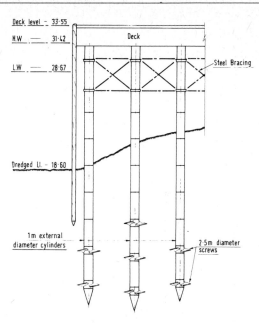

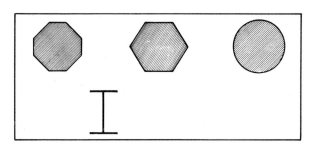

Figure 3.44 Alternative shapes of steel sections for driving to a set.

Figure 3.45 Screw cylinder installation, usually in soft ground.

be taken to a particular set or, by driving beyond the end of the tube, a bulb of concrete can be expanded, and as the tube is withdrawn other bulbs can be formed in the depth of the pile. The bottom bulb can be used as a means of building up a gravel bulb in conditions where piles have been specified to found in the gravel but the gravel is found to be of varying depth giving very thin layers in places over, say, a clay. It is usual to specify a minimum depth of gravel of 1m below the pile base to spread the pile point load into the clay. In Figure 3.42 the tube is used with a detachable flat or pointed end plate and the tube itself is driven using a standard helmet and dolly.

(c) Figure 3.43. This is the taper section pile using standard lengths of spirally corrugated tube with sections progressively increasing in size from the toe which has a solid end plate about 200 mm diameter. Piles of the order of 70 m can be driven with working loads from 60 to 90 tonnes.

(d) Steel piles (Figures 3.44). This type of pile is either of the hollow octagon or hexagon shape or H section and is particularly useful in conditions where the piles are driven over water and the material lends itself to cutting and welding.

(e) Screwed piles (Figure 3.45). This type of pile is not likely to be used in building work.

General References

Skempton, A.W., 'Foundations for high buildings', *Proceedings of ICE,* Part 3.4, 1955.

Terzaghi, K., 'Evaluation of coefficients of subgrade reaction', *Geotechnique,* December 1955.

Baker, A.L.L., *Soil Foundations,* Cement and Concrete Association, 1957.

Proceedings of Symposium on Large Bored Piles. Reinforced Concrete Association, 1966.

BS CP 2004: 1972 Code of Practice for foundations, BSI (under revision).

DOE and CIRIA Piling Development Report, P.G.1, 1977.

Tomlinson, M.J., *Foundation Design and Construction,* Pitman, 1980.

Broms, J., *Precast Piling Practice,* Thomas Telford, 1981.

Chapter 4

Foundation Hazards and Construction Problems

In the chapter on site investigation, groundwater level measurements were described. In addition it is necessary that any natural movement of water should be recognised and contained. Generally, land drains found during construction should be treated with suspicion even when dry, and should be re-routed around the foundations.

4.1.1 Groundwater Table

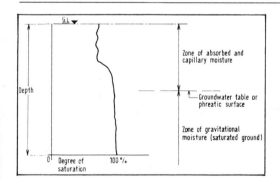

Figure 4.1 Illustration of moisture content of the ground in an exposed area.

The groundwater table is the surface of full saturation of the ground. Above this level water may be absorbed into soil particles as it percolates downwards after heavy rain or may be raised above the water table by capilliarity. Figure 4.1 shows this moisture distribution with a graph of the degree of saturation.

4.1.2 Hydraulic Gradient

The difference in water pressure head at two points represents energy lost through viscous friction as the water flows around the soil particles and through irregular void passages. The rate at which 'head' is lost along any flow passage is called the hydraulic gradient.

4.1.3   Permeability

The capacity of the soil to conduct or discharge water under a given hydraulic gradient is called its permeability. It is similar to electrical conductivity in a solid. A permeable layer is referred to as an aquifer. When the pressure in the aquifer is sufficient to raise the levels in wells or excavations the flow is described as artesian (Figures 4.2 and 4.3).

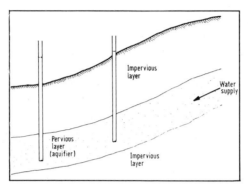

Figure 4.2   Illustration of a hydraulic gradient in permeable soil.

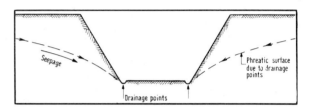

Figure 4.3   Water table drawn down in a cutting by the installation of side drains.

4.1.4   Effect of Water on Bearing Capacity

Granular soil
If the soil or fill is well graded, the voids may be filled with similar strength material and the pressure is carried more easily by avoiding the peak load at the intersurface of the large particles.

If the soil is submerged, the water in the voids is similarly compressed and, if the water beneath a foundation could be contained, it would not affect the load that can be carried without failure. However, the water in the soil beneath foundations over the depth subject to settlement reduces the effective density of the soil by about 50%. The factor of safety is reduced by the same amount and the settlement is nearly doubled. It is normal practice to halve the allowable bearing pressure under a foundation when it is near the water line

Granular soil with large particles
When a granular soil or fill is composed of large particles and is under load, the particles transfer the load by pressure through the places where they are in contact. This condition can arise if water is moving quickly through a graded granular soil. It can leach away the fine grains and, if the moist ground is already at a critical pressure, the crushing of the larger grains at the point of contact can lead to settlement.

Granular soil with finer particles

Silts and fine sands do not suffer in the same way as the above because the forces supported by each grain are small compared with their strength. A particular feature of saturated silts and loose fine sands is that 'spontaneous liquefaction' (soil plus water behaves like a fluid) can be caused by shock. Although this is unlikely in all except machine foundations it is a matter of concern during the construction process (shock might be caused, for example, by pile driving or dynamic compaction).

Cohesive soil

The grains in clay are so small and the movement of the water so slow that, unlike granular soils, they do not suffer bearing capacity reduction.

4.1.5 Effect on Foundation Design

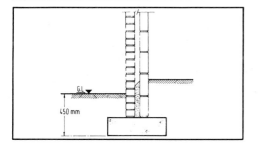

In granular soils where water location is well established it may be possible to use shallow enough foundations to be above the groundwater. In this event frost heave must be considered and BS CP 2004 recommends a minimum foundation depth of 450 mm (Figure 4.4).

Figure 4.4 Foundations should be carried to a depth of 450 mm below ground level in cohesionless soil to avoid frost heave.

4.1.6 Effect of Water Pressure

Pressure in a fluid is equivalent to depth times density and it acts in all directions at a particular level. Figure 4.5 shows the schematic distribution of soil and water pressures on a basement structure in which the building load foundations are also incorporated. It has been known for the walls of basements to be designed for the water pressure while ignoring the upward pressure on the base.

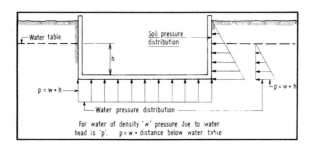

Figure 4.5 Effect of water pressure on a submerged foundation

4.1.7   Sulphate Attack

It is often assumed that soil moisture is simply water, but a number of other materials may be present in dissolved or suspended form. Chemical composition may be important in connection with electrolytic or corrosive effect on steelwork underground, piling or concrete bases both mass or reinforced. Natural sulphates can attack mortar in brickwork foundations and retaining structures. Where there is a hazard – for example, if the soil is pH 6 or less –

Figure 4.6   Proposed cement types and concrete mixes for various levels of sulphate protection. (Crown copyright. Reproduced from Building Research Establishment Digest 240 (250) by permission of the Controller of HM Stationery Office.)

| Concentration of sulphates expressed as SO₃ | | | | Requirements for dense, fully compacted concrete made with 20 mm nominal maximum size aggregates complying with the requirements of BS 882 or BS 1047 | |
| In soil | | in ground water g/L | | | |
Total SO₃ %	SO₃ in 2:1 water: soil extract g/L		Type of cement	Cement* content not less than kg/m³	Free water–cement* ratio not more than
Less than 0·2	Less than 1·0	Less than 0·3	OPC and RHPC BS 12 cements combined with pfa BS 12 cements combined with slag	Plain concrete 250 Reinf. concrete 300	0·70 0·60
0·2–0·5	1·0–1·9	0·3–1·2	OPC and RHPC BS 12 cements combined with pfa BS 12 cements combined with slag	330	0·50
			BS 12 cements combined with minimum 25% or maximum 40% pfa† BS 12 cements combined with minimum 70% or maximum 90% pfa†	310	0·55
			BS 4027 cements (SRPC) BS 4248 cements (SSC)	290	0·55
0·5–1·0	1·9–3·1	1·2–2·5	BS 12 cements combined with minimum 25% or maximum 40% pfa† BS 12 cements combined with minimum 70% or maximum 90% slag	380	0·45
			BS 4037 cements (SRPC) BS 4248 cements (SSC)	330	0·50
1·0–2·0	3·1–5·6	2·5–5·0	BS 4027 cements (SRPC) BS 4248 cements (SSC)	370	0·45
Over 2	Over 5·6	Over 5·0	BS 4027 cement (SRPC) and BS 4248 cement (SSC) with adequate protective coating	370	0·45

*Inclusive of pfa or slag content.

†Values expressed as percentages by mass of total content of cement, pfa and slag.

NOTE 1. Within the limits specified in this table the use of pfa or slag in combination with sulphate resisting Portland cement (SRPC) will not give lower sulphate resistance than combinations with cements to BS 12.

NOTE 2. If much of the sulphate is present as low solubility calcium sulphate, analysis of the basis of a 2:1 water extract may permit a lower classification than that obtained from the extraction of total SO₃. Reference should be made to BRE Current Paper 2/79 for methods of analysis, and to BRE Digests 250 and 222 for interpretation in relation to natural soils and fills, respectively.

BRE Digest recommends special quality bricks to BS 3291 and sulphate resisting mortar 1:0.5:4.5 or stronger preferably using sulphate resisting cement. Good quality well-compacted concrete is often a sound defence but the manner in which it is attacked is important. For example, sulphate crystals in a more or less dry clay will eat away a certain amount of concrete but if water is present, and particularly under pressure, sulphates can be washed from elsewhere in the soil to make repeated attacks. Figure 4.6 shows the accepted protection for proportions of sulphates in both soil and water and Figure 4.7 shows the alternatives for protection when the sulphate content is high.

Figure 4.7 **Alternative methods of foundation protection from sulphates or acids in the soil.**

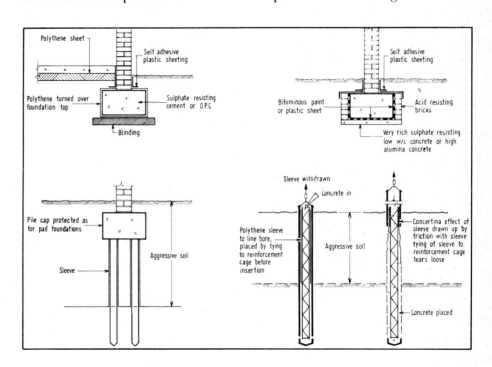

4.1.8 Construction Problems

Deterioration of trenches and foundation excavation is of great concern. If excavation is below the water level, then the method of removing excess water is important.

4.1.9 Pumping

In simple bases carried to just below the waterline it is usual to employ pumps of various types, self-priming being the most effective.

A sump should be provided at a corner of the foundations and maintained by boxing out, for example, with ply sheets and pumping out, until concreting has been carried out above the water line.

If a clay face is being cut up it can be channelled or excavated lower, and granular material used to allow the water to get away quickly to the pump.

4.1.10   Well Pointing

Where a sizeable basement or a large number of foundations are to be constructed in a small area, it is usual to pull down the water table outside the perimeter of the site as indicated for the trench in Figure 4.8. Pumping (or vacuum) continuously for perhaps two or three days may be required to bring down the phreatic surface before excavation. Less vigorous pumping will then maintain this level.

Bleeder wells may be necessary to avoid heave in the excavation. These are pumped like well points but would be sited less densely, as the well points will already have brought down the phreatic surface (Figure 4.9).

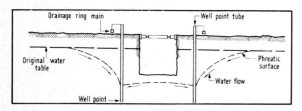

Figure 4.8   Excavation using the well pointing method of lowering the water table.

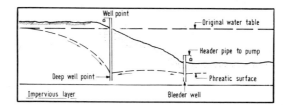

Figure 4.9   Well pointing in stages of depth.

It may be feasible to pump at intervals with suction pumps, but if the soil is such that fine grains will be pulled through the soil towards the pumps, it is necessary to place filters around the suction head or to employ the well-point system; with an adequate distribution of the points, this has the minimum effect on the adjacent soil.

4.1.11   Cut-Off Curtains

Water can be excluded from the construction site by sheet piling, contiguous concrete cylinder piles or diaphragm walls, provided the curtain can be carried sufficiently far into a cohesive soil to avoid percolation under it. Obviously the members providing the cut-off water must be designed for the full water and the retained soil lateral pressure.

4.1.12 Boiling

In Figure 4.10 the foundations were constructed in the gravel without problem. However, as a trench excavated later was deepened, the bottom deteriorated, the flow of water increased and liquid chalk boiled upwards into the trench. (Boiling is a localised bubbling disturbance rather than overall heave.)

Upward flow of water in the chalk had produced pore pressure held in balance by the sand and gravel, but when these were removed the exposed chalk lost strength, allowing water to pressure upwards into the excavation. To deal with this the trench was kept full of water while a deep enough excavation was made, and then the concrete bottom was placed by 'tremie'.

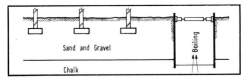

Figure 4.10 Illustration of the phenomenon of boiling.

4.1.13 Concreting in Water

When concrete has to be placed under water, either in a foundation or in a pile, a tremie-pipe is used. With a hopper on top of a tube, the concrete, with high water/cement ratio and high cement/aggragate ratio, is placed in such a way that the tube bottom is always kept below the level of the concrete (Figure 4.11).

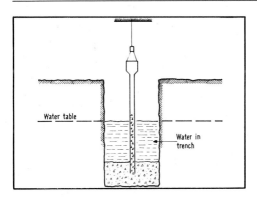

Figure 4.11 Placing concrete by tremie.

4.1.14 Dewatering Precautions

When dewatering takes place the effect on adjacent property must be considered in order to deal with problems indicated previously (see effects on bearing capacity and design of foundations, 4.1.4 and 4.1.5, above).
Similarly, when the water level rises in placed fill, settlement can take place.

4.1.15 Trench/Fill Foundation Problems

This popular method of providing deep (2–3 m) foundations in strip form for lowrise structures can present a hazard in construction. A pattern of soil, say 300 mm topsoil 1 m clayey gravel on top of clay, may, when digging trial pits, stand up for days in dry weather but with heavy rain during construction can

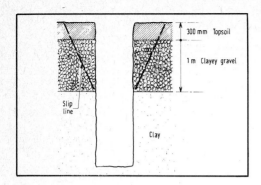

Figure 4.12 Illustration of a possible hazard in trench-fill foundation construction.

collapse due to lubrication of soil particles above the clay and water resting on the top of the clay (Figure 4.12).

4.2 Building on Slopes

In the course of foundation design it may be necessary to actually build into the profile of a slope or prepare ground to create a slope.

Any inclined, exposed ground surface is an unretained earth slope, whether it is nearly vertical or a moderate incline in a single plane or series of planes.

All slopes are subject to natural forces trying to flatten them out; these may take effect imperceptibly or in the form of a landslide. Static equilibrium is maintained through frictional resistance, passive restraining influences on individual particles or cohesion. Equilibrium is disturbed by a temporary increase in forces acting down the slope, or by some decrease in sliding resistance.

The soil can be cohesive (clay), granular, granular mixed with cohesive, or alternatively rock (in the widest sense of the word).

Failures can be rotational slides, translatory slides, or flow slides due to mud flow or liquefaction.

4.2.1 Rotational Slides

The characteristics of a rotational slide are shown in Figure 4.13; the sliding tendency is created by the movement of the mass about the centre of the arc. This type of slide is more likely in a cohesive bank (clay, silt) or an embankment loaded on to the surface of a cohesive soil. Any objects crossing the slip plane will increase the safety factor, i.e. the resistance to sliding. This may be as simple as mass rooting of vegetation or contrived, for example with vibro-compaction (Figure 4.14). The worst slip circle can be found by assuming trial slip surfaces and calculating the safety factor (SF) for each (Figure 4.15). A lower factor may be acceptable where the failure would not endanger a building or life.

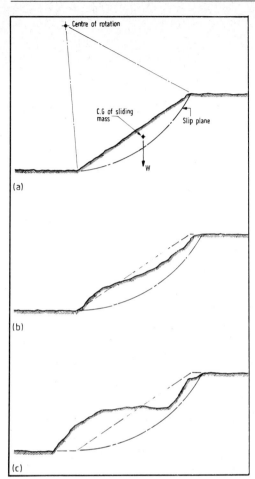

Figure 4.13 Slip arcs in cohesive soils.

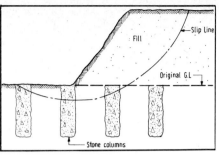

Figure 4.14 Interception of the slip arc by stone vibro compacted columns.

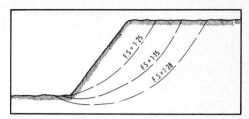

Figure 4.15 Alternative slip arcs constructed in theory to find the lowest factor of safety against sliding.

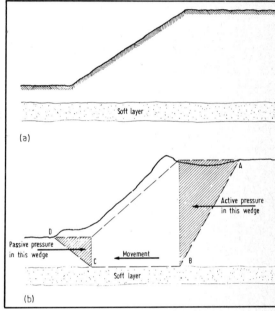

Figure 4.16 (a) and (b) Stages in the formation of a translatory slide on top of a soft layer of soil.

4.2.2 Translatory Slides

The characteristics of a translatory slide are shown in Figure 4.16. This is usually initiated where there is a definite plane of weakness near the base of the slope. The entire area of failure consists of the inclined planes AB and CD as well as the level surface BC on which the block moves.

The location of the horizontal plane of weaknesses is the chief consideration. Natural conditions control the position of this surface to a much greater extent than in the rotational slide; a plane of weakness may be created by a soft layer in an otherwise uniform formation or by a weaker layer in a stratified formation. Finding the area of potential failure depends primarily on a careful subsurface investigation, paying attention to the soil layers which may weaken with time.

4.2.3 Mud Flow and Liquefaction

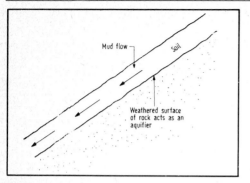

Figure 4.17 Liquefaction.

Although the saturated permeability of a clay soil is extremely low, water can enter a large clay mass at different times of the year. When the clay is not very deep and rests on bedrock, the weathered top surface of the rock serves as an aquifer (a pervious layer) conducting water under pressure from the higher levels to the base of the sliding clay mass. This is referred to as mud flow (Figure 4.17). Flow can only occur in a cohesionless soil when liquefaction occurs. The soil must be fully saturated, in a relatively loose condition and subjected to shock, vibration or shearing strain.

Examination of the slopes by air survey and on the ground and looking at past history will give some guide on flow slides. Drainage design is probably the simplest way of avoiding a further slide.

4.2.4 Natural Slopes

Generally a soil cutting left to its own devices will assume a particular slope – this is called the angle of repose. In cohesionless soils it will be fairly constant for given moisture conditions.

The moisture content is very important and may lead to collapse of the sloping face if there is too low or too high a value. Probable angles of repose are shown in Figure 4.18.

Figure 4.18 Angles of repose for various materials.

Angles of repose for various materials

Material	Angle of repose
Wet clay	15°
Very wet earth	18°
Wet sand	25°
Sandy gravel	26°–27°
Dry earth or dry clay	30°
Damp sand	33°–34°
Dry sand	35°–36°
Shingle	40°
Well-drained clay or moist earth	45°
Clean gravel in natural deposit	50°

4.2.5 Building on Slopes

Building on slopes invariably involves cutting into a slope to create a flat surface on which to build or a combination of cutting and extending the flat surface by fill or a piled platform (Figure 4.19 a,b).

The simplest condition is terraced construction on a moderate slope, which shows the stepping requirement along the slope as defined in the Building Regulations (Figure 4.20). C. & C.A. suggests similar rules for trench fill over

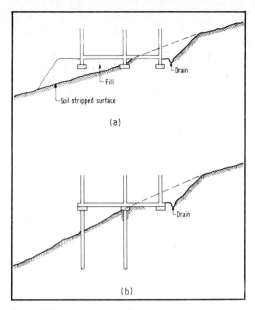

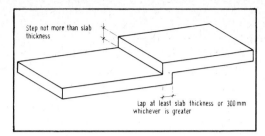

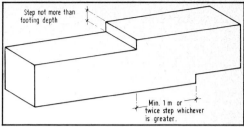

Figures 4.20, 4.21 Overcoming stepping problems in both thin and thick strip footings.

Figure 4.19 Foundation and floor construction on a sloping site.

500 mm (Figure 4.21), though in practice any change in level is likely to be scraped to a curve by the digger bucket.

A deeper step is actually a retaining wall (Figure 4.22), and while in steps up to 1 m the dead weight of the cross wall over may be enough to ensure stability in most soils, its role as a retaining wall must be kept firmly in mind – especially when the DPC passes through the wall and creates a slip plane against lateral pressure and does not effectively transmit tension.

In steps in industrial developments, however, the floor loading on the upper level may be high relative to the dead load on the stepped wall (Figures 4.23, 4.24).

Where houses are built into hillsides two conditions may arise (Figure 4.25) (remember to provide for any necessary under-floor ventilation). In Figure 4.25a the rear wall of the house is a retaining wall and should be specially designed as such; in Figure 4.25b the planning avoids this problem by putting ancillary uses on the lower level and living space on the upper level.

One alternative to the above construction where the plan permits is to use a raft, half cutting and half filling the site, or fully filling or stepping (Figure 4.26). In industrial work it may be appropriate to suspend part of the ground floor and either fill under or use for open storage for parking in the open wedge (Figure 4.27).

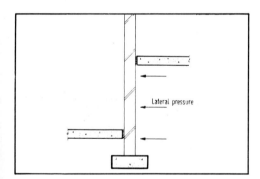

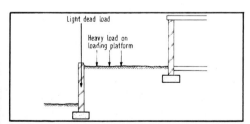

Figures 4.22, 4.23, 4.24 Effect of stepped floors on foundation.

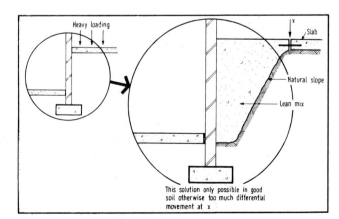

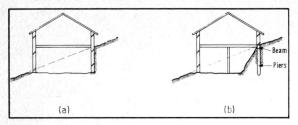

Figure 4.25 (a,b)   House built into a hillside.

Figure 4.26 (a,b,c)   Alternative raft forms in a sloping site.

Figure 4.27   Industrial building on a sloping site with use made of the underneath void.

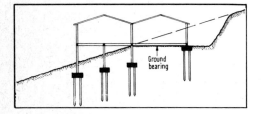

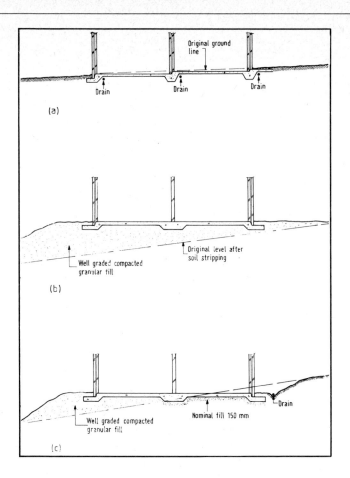

4.3   Foundations in Poor Ground

Poor ground with low bearing capacity or filled sites is increasingly being used for building as efforts are made to limit incursions into the Green Belt. Fill, silt, peat and very soft clays are the materials under consideration and generally an engineer will be involved in interpretation of site investigations and the foundation design. Nevertheless, architects too often put too much money into the ground and an examination of the options suggested hereafter may lead to a useful dialogue between architect and engineer.

Ground conditions may vary enormously, some requiring foundations to be strengthened a little above normal, e.g. strip foundations heavily reinforced, to a point where the soil is so poor that it is prohibitively expensive for building. The extent of foundation work hinges in part on how much risk the client is prepared to agree to in writing (informal agreements are no protection against future litigation). Much fill and some other poor ground varies in quality, so that foundation performance is more that usually unpredictable. Risks may sometimes be reduced by taking more elaborate precautions, such as piling rather than using a stiff raft. On some commercial projects a client may agree to less elaborate foundation work to keep down capital cost. Site investigations will involve more frequent boreholes and pits, especially on fill since this is likely to be of variable composition. It is also important to define the boundaries of poor ground if this is localised on the site.

For fill, its history may be clarified from map sources from local authorities, local residents and past owners of the site.

4.3.1   Natural Poor Ground

Soft clay
Clay of low bearing capacity is not especially problematical, but settlement and differential settlement assume greater importance and the clay is more susceptible to seasonal changes.

Silts
Silts derive from original rock types by a mechanical break-down process and are mainly composed of fine quartz material. However, they are often found in river, lake and marine deposits. In rivers the silt is deposited in the flood plane, sometimes on the terraces bordering the flood planes. The silt can be interlayed with sand and gravel. Estuaries and deltas have similar silt deposits. Silt is carried into lakes where it forms in bands and can be associated with varved clays which contain particles similar in size to those of the silt. Silts can form as the result of blowing by the wind, in which case the particles form up in layers of uniform size.

Porosity, void ratios and density are influenced by the grain-size distribution. Consolidation is influenced by grain size, natural water content and porosity.

Peat
Peat derives from decomposed and disintegrated remains of plants fossilised under conditions of high water content and incomplete aeration.

Present surface deposits of peat have originated since the ice age; buried peats are much older.

Peats have also grown in marshes and lakes since the ice age; in which case they can be interlayered with silt. The well-known deposits in the Fens have developed during changes in the level of the sea since the ice age. Deposits are found mainly south of the Wash and in Suffolk. Perhaps the best recognised peat is that found in bogs on the moors.

The water in the peat is largely 'free water', usually acidic (ph values 5·5–6·5) but occasionally neutral or even alkaline. Differential and excessive settlement is the main design problem in peat conditions. A loaded soil is normally 'contained' by the adjacent soil, but in the case of peat there is little lateral resistance; consequently high shearing stresses are induced by comparatively light loads and when the settlement reaches a certain point there can be both creep and lateral spread. This may cause rotational slips and upheaving of the adjacent ground.

As the load is applied the free water is squeezed out so that not only is the settlement considerable but it occurs quickly. When a certain amount of water has been expelled the peat has a higher bearing value. The voids ratio is of the order of 10:15 so that complete consolidation means a reduction of $\frac{1}{10}$–$\frac{1}{15}$ of the original volume.

4.3.2. Filled Ground

Filled characteristics
The term fill covers very diverse materials from natural soil and rock to domestic and industrial wastes, sometimes of great depth. Fill may be well graded inert material compacted in thin layers and providing acceptable building land. Fills are not inherently poor ground, but they are all suspect.

Controlled fill
Controlled fill may take place with building in mind. By careful preparation of the surface to be filled, diversion of ditches, drains, ponds, etc., and by building up layers of fill compacted by specified equipment, it is possible to build as soon as the levels have been achieved.

Great care is needed with fine grained material such as clay. Clay fill on the average building site is unlikely to be successful. However, the specification in architect-controlled projects frequently offers no real control of fill and the cheap process – return, fill and ram – covers a multitude of sins.

Uncontrolled fill
There are four main problems, all of which could occur on one site: large general settlement; differential settlement; collapse settlement; chemical attack.

Large general settlement
Newly placed uncompacted fill will settle substantially under its own weight, indeed this settlement may be more than that due to light building such as housing. The graph (Figure 4.28) shows fill settlement under its own weight, most of which has taken place in the first two years. Refuse rather than soil is a hazard, largely for the reasons noted below for differential settlement. Wet cohesive soils, even if well compacted, can be slow to settle because water will be slow to dissipate.

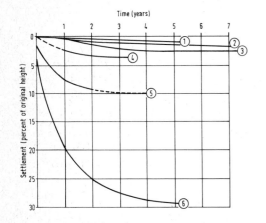

| Description of curves | | |
No.	State of compaction	Material
1	Well-compacted	Well-graded soil
2	Medium	Rock fill
3	Lightly-compacted	Clay and chalk
4	Uncompacted	Sand
5	Uncompacted	Clay
6	Well-compacted	Mixed refuse

Figure 4.28   Curves showing the settlement of different fill materials with time, after Meyerhof.

Differential settlement

Variable ground is likely to settle differentially. The problems are illustrated in Figure 4.29 (they also apply to variable natural ground). Differential settlement may be due to varying depth of fill and varying composition, especially parts of fill that decay such as organic matter and metal containers.

Figures 4.29 (a,b,c,d) Settlement and differential settlement problems in fill.

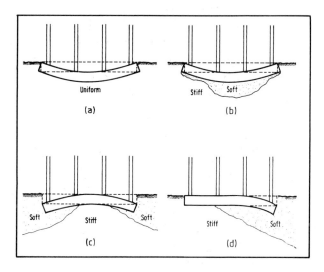

Collapse settlement

Loose unsaturated fill materials are usually liable to collapse compression and hence settlement on inundation with water (a change in moisture content or water table level can also greatly affect rates of decay or decomposition of refuse). An attempt at inundation via surface trenches by the BRE was not particularly successful. But on occasion more general inundation of the site may be tried on deep fills.

Chemical attack

The range of potential chemicals and their concentrations cannot be conveniently summarised. Tests must be done to assess risk to concrete, metal and mortar of foundations, and to assess health hazards. Decomposition of organic matter may result in potentially hazardous concentrations of methane or carbon dioxide.

Other chemicals present potential hazards to human and plant health. A large proportion of combustible material in a fill into which air can penetrate involves risk of underground fires which may not be evident on the surface. Where fires have occurred, the support offered by this ground is likely to be variable and very risky to build on. Some contaminated fill materials may be deemed as notifiable wastes under the Control of Pollution Act 1974 and are subject to control if removed from site. The local Health and Safety Executive or the Control Unit on Environmental Pollution of the DOE can give advice. Note that piling, especially if drilled to provide open bores before concreting or stone columns (see below) can provide paths for chemicals at different levels in fill to mix, possibly with explosive results.

Defeating chemical attack

Surprisingly perhaps, polythene sheet is an effective barrier against very many chemicals and can be laid beneath rafts or wrapped around footings. Precautions may be relatively straightforward as protection against severe attack. Driven piles *in situ* may be sleeved through aggressive soils, though this is not easy. Polythene sheet sleeves tied to the reinforcement cage and placed with it tend to tear at the ties, come loose and concertina upwards when the outer sleeve is withdrawn. A tube of galvanised corrugated sheet steel or rigid pvc inside the outer sleeve will stay rigid and can with some difficulty be held down during withdrawal of the outer sleeve. Driven piles are more easily treated by prior surface coating, e.g. with pitch epoxy. They also have the advantage of the higher quality, denser concrete, achievable in the factory.

4.3.3 Alternative Methods of Founding in Poor Ground

The obvious solution, if the area of the site permits, is to place the buildings on the better ground and leave the poor ground for single-storey garages, play areas, landscaping, etc. If the buildings must of necessity cover the bad ground then there are several options.

(a)   Remove the bad ground and replace by good fill, obviously only practicable where the fill is not of great depth.

(b)   Improve the ground condition by surcharging it, later removing the excess surcharge material. This is only practicable where time is not the essence of the project, a rare event.

(c)   Improve the ground condition by vibration technique or dynamic compaction.

(d)   Pile through the fill to firm soil.

(e)   Use stiff rafts or the combination of raft and piles.

Rafts

Flexible rafts (i.e. thin rafts) accept more movement and so require a relatively flexible structure, for example, CLASP schools. Sheet cladding on a steel frame is suitable, though infill of brick or block is possible with the usual expansion joints and preferably weak mortar, 1:2:9 if possible. If loads are spread evenly over the slab, less flexural demands are made on rafts, with less risk of differential settlement. A raft often appropriate for Fenland peat, would suit light structures. For much heavier structures the beam element of the raft will increase until in the cellular form it is effectively a beam of one depth.

Stiff rafts for relatively small buildings can be substantial. Figure 4.30 shows a proposed design of a raft for two-storey houses on a filled quarry (23 m deep) site. Dynamic compaction was chosen for the consolidation of the upper 6 m of the fill. The design assumes a total settlement of 100 mm and a differential of 60 mm.

Figure 4.30 Raft to absorb considerable differential and total settlement on a filled site.

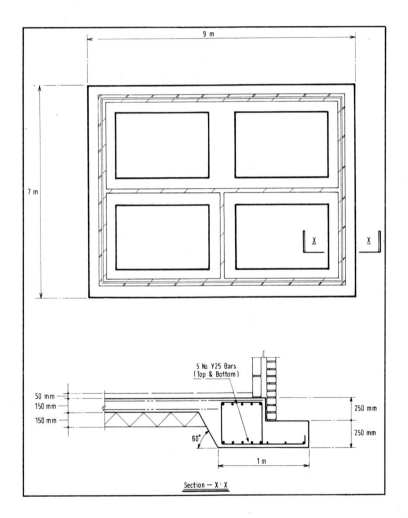

Deep silts require more robust rafts, and occasionally buoyancy rafts are used; these are cellular to ensure that the weight of the raft is less than the weight of the displaced silt, thereby producing passive uplift. The cells should preferably be expanded polystyrene to avoid water penetration which would defeat the object. Alternatively, useful space can be provided by using a basement as the

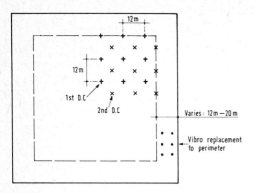

Figure 4.31   Layout of the site at Thorpe where the fill was consolidated by dynamic consolidation.

buoyancy chamber to reduce total downward load, which is carried on piles penetrating the silt to a stiffer layer.

Dynamic consolidation

Dynamic consolidation involves dropping a very large weight from a considerable height, e.g. 10–20 tonnes from 15 to 25 m (Figures 4.31, 4.32). It is a technique requiring very large equipment to be used on large sites. The plan of a site at Thorpe, Surrey, shows a pattern of dynamic consolidation for 110 kN/m². Note the 20 m space between consolidation area and site boundary to allow for safety of adjacent buildings and services. Even with this there are sometimes complaints of vibration and concern from neighbours about damage, though in fact any damage is rare. Vibration is certainly felt, but cracks

Figure 4.32 (a, b)   Dynamic consolidation in action.

(a)

(b)

in buildings may not be new, just newly noticed. (A similar reaction is often produced by pile driving.) An extra cost of dynamic consolidation may be the need to remove old foundations, sewers, etc., beneath fill, to a depth of a few metres. Dynamic consolidation is particularly effective in granular material but can be applied to mixed ground. In low permeability silty soils, high water pressures develop as a result of tamping stresses, which promote water dissipation faster than under static loading. Trenches may be required to take away water and the craters must not be allowed to fill up with water, otherwise the dynamic effect of dropping the weight will largely be lost on impact.

In every individual application of dynamic consolidation, drop pattern, energy level and number and phasing of tamping passes are specially chosen. Three or four drops are made in one position, creating a crater sometimes 2·5 m deep and 5 m in diameter. As the crater section shows, cores of compacted soil are developed with an apex angle of between 20° and 40°. The craters are filled with good compacted material as the drop pattern is developed. The final ground level may finish 200–1,500 mm lower.

Piling

If a pattern of piles is needed to provide support from lower, better soils to a floor, it will be more economical to design the floor slab coned down to the pile head to avoid pile cap and beam costs, which can be large. Depending on the nature of the ground it may be appropriate to support the structure on piles and allow the floor slab to settle independently. Provided the slab is designed with construction and contraction joints properly dowelled, the individual panels, of say 6 × 4·5 m, can articulate and take up the general dishing nature which the fill is likely to assume (Figures 4.33, 4.34). Compare with usual arrangement (Figure 4.35).

Figure 4.33   Articulated connection of a pile tie to avoid damage by settlement of the floor over.

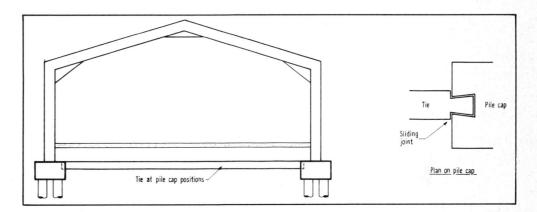

Figure 4.34   Detail arrangement of tie in Figure 4.33.

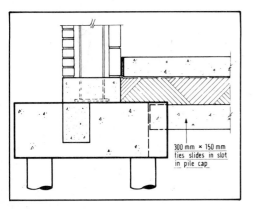

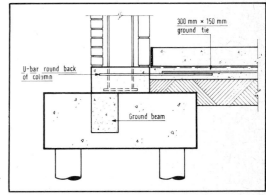

Figure 4.35   Pile tie beam in industrial building where normal settlement is predicted.

Vibration technique

Compaction or replacement methods of vibration can be used to build up stone columns – a combination of piling and ground improvement. The vibrator is like a large poker vibrator fitted with jets at the tip for compressed air or water. Typically the vibrator weighs 2 tonnes and is 5 m long, though it can be lengthened with extension tubes (Figures 4.36, 4.37).

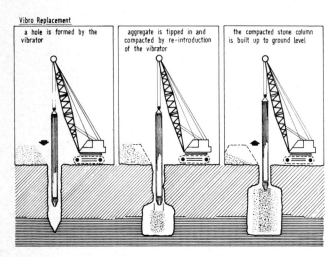

Figure 4.36   Stages in the process of vibro replacement.

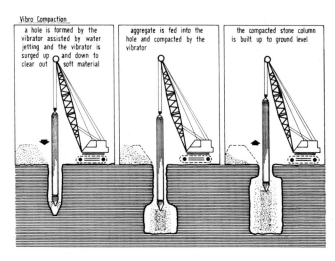

Figure 4.37   Stages in the process of vibro compaction.

For more granular ground the replacement method involves jetting with compressed air at the head of the vibrator to lead into the ground. Granular material is tipped into the hole in layers and compacted by the vibrator. For more cohesive or silty soils, the compaction method involves surging the vibrator up and down the bore to allow the jetting action of water to remove softer material. Aggregate is tipped down the side of the vibrator and compacted. These methods build up stone columns to ground level. However, these are not independent columns. Vibration pushes the stones into surrounding ground, compacting and strengthening it. As Figure 4.38 shows, cohesive ground is more easily penetrated, its cohesive strength supporting the column. Testing can be carried out fairly cheaply by jacking up the machine (of known weight) on top of a stone column and measuring settlement. Stone columns can act like piles connected by lightly reinforced beams or, for more concentrated structural column loads, several stone columns can be capped like a pile group and the structural column built off the top. Floor slabs may be supported by placing stone columns over the site.

Up to 25 stone columns 6 m high can be produced by one machine in a day, so vibration methods can be used on relatively small sites.

Figure 4.38   Extent of the effect of the vibro technique in
different soils.

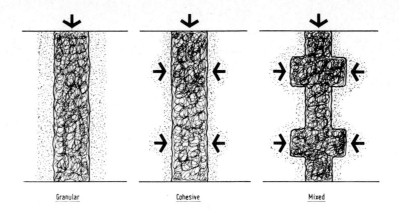

Granular   Cohesive   Mixed

4.3.4   Designing for the Movement

There are limited applications of articulation in residential buildings having regard to the type and quality of the desired finishes. More scope is available in buildings such as schools, factories and warehouses. The CLASP form of construction uses a flexible raft and an ingenious spring-loaded bracing system in the frame which accommodates movement while not overstressing the frame and at the same time still providing stability to the frame. Sheet materials are used in preference to masonry for the cladding, hanging from the frame as opposed to ground support.

If a thin raft is used and the loads are spread evenly, less flexural demand is made on the raft and there is less risk of differential settlement. A raft often used in Fenland peat areas (Figure 4.39) would be appropriate for light structures. For much heavier structures the beam element of the raft will increase until in the cellular form it is effectively a beam of one depth.

Figure 4.39   Typical raft used in the Fenlands.

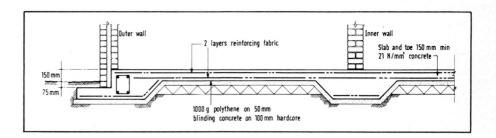

4.3.5 Buoyant Foundations

There are parts of the world where silt extends to enormous depths and piles of the order of 60 m length might be required. Having regard to the moisture content of the silt it is possible to construct a hollow, or, preferably, expanded polystyrene-filled raft in which the combination of low bearing pressure and buoyancy in the silt can equal the applied load. Some movement must occur as the building is constructed and if superimposed load is variable, due account must be taken of possible tilting and connection of services outside the raft area (Figure 4.40).

Figure 4.40   Cellular raft.

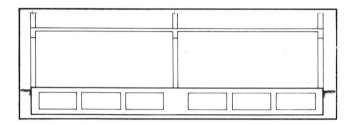

4.4   Ground Movement

Settlement of the soil beneath foundations described in Chapter 2 results from the change in pressure at the founding level.

There are, however, movements of the soil which may be active, that is to say the soil moves the foundation rather than the foundation moves the soil.

This can result from varying moisture and temperature conditions, shock and vibration and the effect of removing layers of minerals at various depths.

4.4.1   Clay Soils Subject to Varying Moisture Conditions

Few readers will be unaware of the panic caused by the dry summers of 1975 and 1976 in the UK when drying out of the clay caused the settlement of many house foundations and the following wet winter produced a swelling of the clay causing equal damage not only to walls but to ground-bearing slabs and internal partitions resting on them. While some clays of the fissured type show this marked movement due to moisture change, all clays are affected by the roots of growing trees and the drying roots of trees which have been cut down.

Particular movement occurs in the bloating clays of the type used in brick-making where in a treed area desiccation can take place to a depth of 2–3 m. There is no simple field or laboratory test to identify the potential shrinkable clay, but clays with a liquid limit of more than 50% may be troublesome. This is not a positive criterion as water content must be considered at the same time. If the measured natural water content of the clay is significantly wetter than the soil's plastic limit, then the test materials are not in a desiccated stage. Furthermore, the presence of ground water can allow the plant growth to obtain its necessary moisture from the water table rather than having to overcome pore/capillary attraction within the soil (Tables 4.1, 4.2).

Table 4.1 Clay shrinkage potential, after BRE.

Plasticity index %	Clay fraction %	Shrinkage potential
> 35	> 95	Very high
22–48	60–95	High
12–32	30–60	Medium
< 18	< 30	Low

Table 4.2 Shrinkage potential of some common clays, after BRE.

Clay type	Plasticity index %	Clay fraction %	Shrinkage potential
London	28	65	Medium/high
London	52	60	High
Weald	43	62	High
Kimmeridge	53	67	High/very high
Boulder	32	—	Medium
Oxford	41	56	High
Reading	72	—	Very high
Gault	60	59	Very high
Gault	68	69	Very high
Lower Lias	31	—	Medium
Clay silt	11	19	Low

Where the adding of moisture to the clay causes swelling, e.g. where the roots of a dead tree no longer take moisture from the clay, the resulting heave can affect foundations and ground-floor slabs, and in certain circumstances can produce a lateral pressure on the foundation in the ground, e.g. a brickwall on a strip foundation or a trench concrete fill foundation.

These heave pressures can be avoided by using resilient or collapsible materials between the structure and the clay.

Trench concrete fill foundations, considered to be an economical solution up to a width of about 900 mm, were considered suspect when placed in the type of clay now being considered as the concrete could take adhesion when first placed but the adhesion could later be lost when drying out of the clay in the top metre occurred. This has led to demands for polythene to be placed on the sides of the trench and expanded polystyrene or clay board on the inside of the trench; this creates enough problems to make the system unworkable in many situations. Clays can also move due to drying out. As described above, this can be an effect of air temperature but can also be the result of applied heat in the use of the

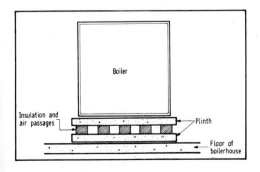

Figure 4.41 Avoiding damage to a boiler house floor.

building. The floor of a boiler house or a brick kiln are particular cases in point (Figure 4.41).

4.4.2 Soils Subject to Frost

Frost conditions vary immensely in different parts of the world. In the UK a month of temperatures below freezing would be considered as abnormal but in arctic and antarctic regions there is permanently frozen ground – permafrost – varying from a few metres to enormous depths. Permafrost areas occur in Canada, Alaska, Greenland and in as much as 50% of Russia.

The UK conditions affect chalk, chalky soils and silty soils. Clays and granular soils where the frozen water has somewhere to expand are not affected. Foundations placed at 450 mm below ground level are unlikely to be affected but much greater depths are needed in countries such as Canada and Scandinavia which have much colder winters.

Building in permafrost areas is faced with many varied problems (Figure 4.42).

Figure 4.42   Typical house foundation in a permafrost area.

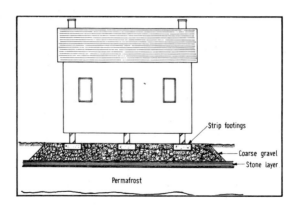

Strip footings
Coarse gravel
Stone layer
Permafrost

4.4.3   Moisture Leakage into Soils

One of the first considerations when examining a damaged structure is to look for leaking drains, gulleys, etc. Leakage may take out the fine particles of a granular soil or cause a clay soil to swell. Water mains, surface water and soil drains, even badly connected wastes, can all over a long period cause damaging settlement or heave.

4.4.4   Removal of Minerals

Removal of coal and salt from shallow or deep deposits constitutes a considerable hazard in those parts of the world where these deposits are found. Two

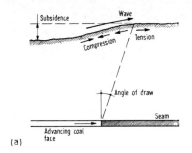

(a)

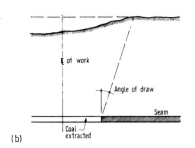

(b)

Figure 4.43 Movement in the ground in long wall mining.

conditions arise, one building in an area where workings have taken place and the other in an area where there are to be future workings.

In the latter case mining experts can give a reasonable picture of the manner in which the ground will deform during the period of working and thereafter, having regard say in coal mining to the depth of the seam from the surface, the thickness of the seam, the method of refilling and the general speed of the operation. There is also the possibility of buying the mineral beneath the building and therefore avoiding movement, but the political situation, certainly in the UK, is rapidly preventing this happening.

A number of publications describe in some detail the different methods of working and precautions which can be taken to prevent or minimise damage to structures. Put simply, the working produces a wave in the ground which the building must ride (Figure 4.43).

Initially the ground under the building is in compression and later moves into tension. If the building is flexible and the ground can slide underneath it, then a minimum of damage occurs. This is not too difficult to achieve in a dwelling where the structure can be carried on a smooth-bottomed raft. The raft, when laid on polythene and a thick sand bed, stands a reasonable chance of the ground sliding underneath it. If a trench can be dug around the building deeper than the sand bed and filled with large particle gravel, then a further protection is provided. Deep foundations such as piers and piles are vulnerable to the movement as the ties just below surface level will tend to either push or pull the tops of the foundations.

The *NCB Handbook* suggests that the shallow raft foundation is the best protection but this is not applicable to industrial buildings even when these latter can be divided into smaller units. Structures should be massively rigid or articulated, the latter course being the only economical way of dealing with a factory or warehouse.

Various solutions have been put forward such as three-point suspension or jacking facilities from the foundation, which is allowed to move.

In some cases it is possible to fill or inject the mine passages, probably the solution to apply to old workings where predictions of ground movement either with or without the loading of the structure may be impossible.

4.4.5 Disused Shafts

A possible hazard in areas underlain by chalk is that wells may have been excavated in the past and capped over when becoming disused. A foundation sited within 45° of the rock chalk level in such a well will be in danger if the well is not filled with fully consolidated hardcore, which may be difficult to achieve, in which case a weak concrete may be placed.

An old mine shaft may have been capped off with a concrete slab and could be treated in a similar manner.

4.4.6 Vibration

Vibration can result from machinery or from transport, particularly in tunnels. The effect of the vibration may be to compact a granular soil and cause settlement beneath other foundations or to transmit noise and vibration into office or residential accommodation. Assuming that the soil does not settle, a simple criterion in buildings is that if the human beings can stand the vibration so can the building. Nevertheless, a rumble at 1 a.m. or a clinking of glasses on the table can be extremely irritating.

Underground railways

The first construction of the London Underground Railways was on the cut-and-cover method, often lying under or alongside the mainroads. Later constructions went deep into the London Clay, employing tunnelling machines. In those built since 1960 due regard has been shown for existing property and the track has been laid on rubber mountings. Buildings with basements have been constructed over the top of or adjacent to the tunnels using rubber springs between the columns and the foundations.

Figure 4.44 shows the relationship of a West London housing development and the extension line to Heathrow Airport. A simple stethoscope applied to a concrete ground slab detected the noise of the trains in the nearest housing unit and this was protected in the manner shown in Figure 4.45, with a proprietary

Figures 4.44 **Plan of flats adjacent to a West London tube**
and 4.45 **station and detail of the antivibration system.**

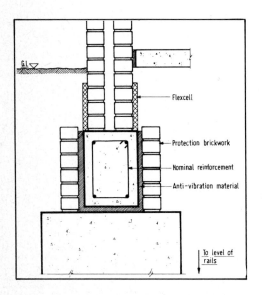

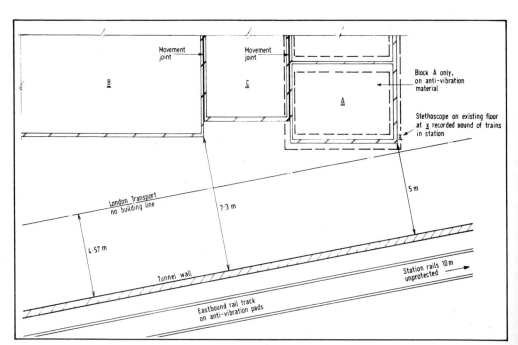

rubber carpet mounting surrounding a foundation beam entirely isolating it from the adjacent ground and mass base.

Modest mechanical plant can have its vibration damped out by anti-vibration cup mountings but the heavier plant such as drop-hammers requires very special treatment.

4.4.7 Earthquakes

An earthquake is a natural subterranean shock giving rise to waves of elastic compression, which radiate from the centrum in all directions as concentric spheres. The point where the waves first meet the surface is the epicentrum from which they move outwards in ever widening circles, like those from a stone dropped in water. At the epicentrum the movement is up and down: further away it is in the direction of a line joining the place in question to the centrum. The forces arising can be applied in any direction as the ground heaves up and down and tears apart. Earthquakes are measured as a number on the Richter scale and on the basis of level of damage, those areas of the world affected tend to have imperical numbers of 'g' forces as statutory design loads of buildings. That is to say that if the figure is '1' it must be assumed that the weight of the building can act upon itself in any direction. All the problems of mining previously described apply, except that the earthquake happens in seconds rather than in weeks. Figure 4.46 shows the areas of the world most likely to be affected by earthquakes.

Figure 4.46 The earthquake areas of the world where considerable damage has occurred.

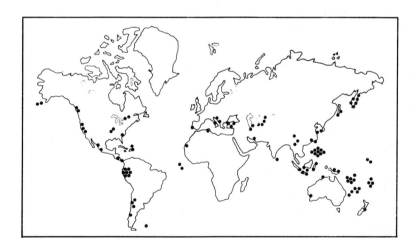

General References

BRE Digest 250, Concrete in sulphate-bearing soils and groundwaters.

BRE Digest 274, Parts 1 and 2, Fill.

ICRCL 17/78 Notes on the redevelopment of landfill sites. DOE.

ICRCL 61/84 Notes on the fire hazards of contaminated land. DOE.

ICRCL 59/83 Guidance on the assessment and redevelopment of contaminated land. DOE.

Lacey, W.D. and Swain, H.T., 'Design for mining subsidence', *Architects' Journal*, 24 October 1957.

Concrete Corrosion – Concrete Protection. 1. Biczok, Hungarian Academy of Sciences, 1964.

BS CP 2004: 1972 Code of Practice for foundations, BSI (under revision).

Wiegel, R.L., *Earthquake Engineering*, Prentice-Hall, 1970.

Subsidence Engineer's Handbook, National Coal Board, 1975.

Chapter 5

Underpinning

Underpinning is a method of refounding a building, usually because the original foundations have been placed in poor soil or alternatively the soil has been subjected to shrinking or swelling due to heat and moisture or due to root action.

An alternative reason for underpinning may be the construction of an adjacent structure with a basement or foundations at a lower level.

There are a variety of methods available varying from complete lowering of the foundations on a continuous footing to beam and stool arrangements, some of which are patented, to piling and injection of chemicals.

The underpinning industry in the UK has been very active since the 1975–6 drought, and the dry period of early 1983–4 followed by the actual drought of 1984 will undoubtedly lead to a new cycle of underpinning.

A great deal of this work relates to two- and three-storey houses in the south-east which have foundations of negligible depths. In these cases the fully underpinned trench foundation is to be preferred as it produces a perimeter wall, say, 2–2·5 m below ground level which retains a fairly constant condition of the soil within the building confine and avoids the necessity of underpinning the internal walls. This does not mean that the occupants see all the work done from the outside, as in semi-detached and terrace houses the party wall usually subsides as well and has to be dealt with from one side or the other. Houses with basements are a particular hazard as there is a large differential settlement between the lower founded basement bottom and the rest of the building, usually leading to a large crack at one or the other end of the basement which is wide at the top of the building and peters out at basement level.

Figures 5.1 and 5.2 show typical underpinning procedures and Figure 5.3 and the typical specification (pp.84, 85) shows how to deal with a situation where the starting and founding levels of the underpinning are not known and drains and footpaths around the property have to be reinstated.

Where more sensitive structures are involved it may be necessary to use piles (Figures 5.4–5.6) carried to a much firmer lower stratum and to jack the

Figure 5.1   Standard underpinning procedure.

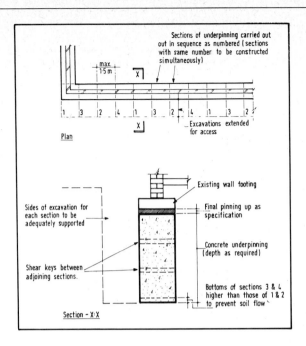

Sections of underpinning carried out in sequence as numbered (sections with same number to be constructed simultaneously)

max. 1·5 m

X

Plan

Excavations extended for access

Existing wall footing

Sides of excavation for each section to be adequately supported

Final pinning up as specification

Concrete underpinning (depth as required)

Shear keys between adjoining sections.

Bottoms of sections 3 & 4 higher than those of 1 & 2 to prevent soil flow

Section – X·X

Figure 5.2   Beam and pad underpinning method.

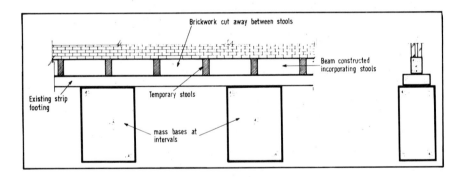

Brickwork cut away between stools

Beam constructed incorporating stools

Existing strip footing

Temporary stools

mass bases at intervals

structure load into the piles so that the initial compression of the piles and soil when accepting the load do not lead to a movement in the building. Obviously the building load must be assessed very accurately or the building may be damaged by the jacking.

One particular method of improving the foundations under a structure, often a bridge foundation, is to bore small diameter holes and employ piles of the Palli Radicce type (Figure 5.7).

Wherever underpinning is employed it must obviously take the structure load to a far stronger, less compressible stratum.

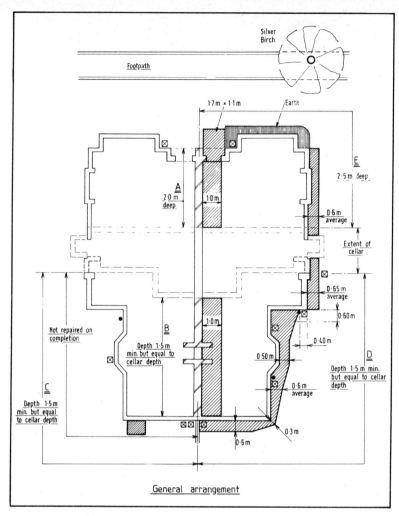

Silver Birch

Footpath

1·7m × 1·1m

Earth

E

2·5 m deep

A

2·0 m deep

1·0 m

0·6 m average

Extent of cellar

Not repaired on completion

0·65 m average

B

1·0 m

0·60 m

Depth 1·5 m min. but equal to cellar depth

0·40 m

D

0·50 m

Depth 1·5 m min. but equal to cellar depth

C

Depth 1·5 m min. but equal to cellar depth

0·6 m average

0·3 m

0·6 m

General arrangement

Figure 5.3 Record drawing of a pair of semi-detached houses underpinned by the continuous support method.

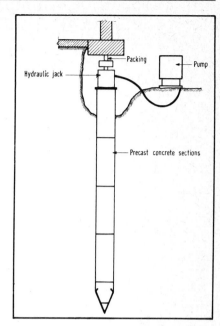

Packing

Pump

Hydraulic jack

Precast concrete sections

Figure 5.4 Jacking pile to underpin a sensitive structure.

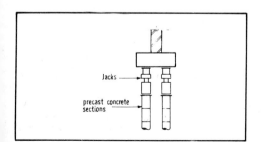

Jacks

precast concrete sections

Figure 5.5 Jacking piles where both sides of the structure are accessible.

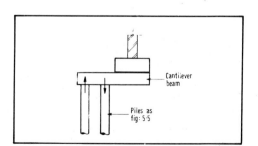

Cantilever beam

Piles as fig: 5·5

Figure 5.6 Jacking piles where the foundation cannot be undermined and the load to the piles is eccentric.

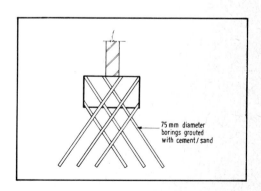

75 mm diameter borings grouted with cement / sand

Figure 5.7 The small bored pile method of underpinning.

SPECIFICATION FOR THE UNDERPINNING WORKS AT 13/15 OAKHURST ROAD NW10

(1)   The underpinning works required in the above property are defined on Drawing No. 8327/2 and shall be carried out in accordance with this Specification.

(2)   The properties are in different ownership and consequently different sections of the underpinning are defined on the Drawings as Units A-E inclusive and it will be necessary to price these items separately so that the cost of the works can be apportioned to the appropriate Client.

(3)   The depth of existing foundations and levels at which a suitable clay base will be found have been assumed and the brief Schedule of Quantities makes provision for variation.

(4)   The bottoms of all excavations are to be taken down to the levels required by the drawings or other instructions and are to be to the satisfaction of the Engineer and/or District Surveyor.

(5)   All excavated material not required for back filling shall be removed to a tip to be provided by the Contractor.

(6)   The sides of excavations shall be properly supported and retained by good sound timbering (or by other satisfactory methods) wherever necessary. All such timbers shall be carefully removed as the excavations are filled in. The removal of the timbering as aforesaid shall be done in such a manner as not to endanger the works and shall not relieve the Contractor of the responsibility for ensuring the stability of the works.

(7)   The bottoms of all excavations shall be carefully trimmed and finished to the specified levels and all loose material removed.

(8)   The Contractor shall ensure that prepared formations are not damaged by weathering. Concrete shall be placed on the same day the excavation has taken place unless the formation is blinded with concrete or otherwise suitably protected.

(9)   The Contractor shall be responsible for keeping all surface water away from the new foundation but provision is made in the brief Bill of Quantities for pumping any water which arises from the condition of the soil below ground level.

The Contractor's attention is drawn to the necessity of providing adequate means of supporting, shoring or upholding by any other means adjoining structures, foundations, roads, services, etc. , during operation on the site.

(10)   The Contractor shall have access from the ground floor of No.13 to underpin the party wall, Units A and B. The remaining work shall be carried out from the external perimeter of both houses.

(11)   In situ concrete work shall be entirely in accordance with Code of Practice CP 114 and the London Building Bye Laws as may apply to the satisfaction of the Engineer and the District Surveyor. The concrete shall be a mix 1:2:4 with a minimum cement content of 280 kg cement per cubic metre of finished concrete with a maximum size aggregate of 40 mm and a maximum water cement ration of 0·6.

(12)   The concrete shall be placed in the excavation and thoroughly compacted for the underpinning and brought to a level 75 mm below the existing foundation.

(13)   After the in situ concrete has been in position for a minimum of 12 hours the 75 mm gap shall be filled by dry packing a small aggregate concrete which shall be of a mix not less than 1:3 by volume and shall be of such a consistency that it just remains in the form of a ball when squeezed between the hands.

(14)   The extent of underpinning shall be restricted to alternate lengths of the perimeter to ensure the support of the building by the remaining piers of soil.

(15)   The concrete in the underpinning shall be not less than two days old before the adjacent wall pier is removed.

(16)   The Contractor shall be provided with a source of water and electricity from House 13.

(17)   On completion of the works the Contractor shall leave the site in a clean and tidy condition.

```
                BRIEF BILL OF QUANTITIES
                          for
       THE UNDERPINNING WORKS AT 13/15 OAKHURST ROAD NW10
                                              £      p

1.  Lump Sum to underpin Wall A entirely in accordance
    with the Specification and Drawing No. 8327/2........

2.  Ditto ditto        Wall B  ditto ditto    ........

3.  Ditto ditto        Wall C  ditto ditto    ........

4.  Ditto ditto        Wall D  ditto ditto    ........

5.  Ditto ditto        Wall E  ditto ditto    ........
```

Provisional Items

```
6.  Provisional Item to add or subtract if the existing
    base is not at the depth shown on Drawing No. 8327/2
    per 100 mm depth per m run.

    Wall A ..........................................

     "  B ..........................................

     "  C ..........................................

     "  D ..........................................

     "  E ..........................................

7.  Provisional Item to add or subtract if the bottom levels
    of underpinning are modified at the instructions of the
    Client per 100 mm depth per m run.

    Wall A ..............................................

     "  B ..............................................

     "  C ..............................................

     "  D ..............................................

     "  E ..............................................

8.  Ditto ditto if base level is undercut in the clay per m run.

    Wall A ..............................................

     "  B ..............................................

     "  C ..............................................

     "  D ..............................................

     "  E ..............................................

9.  Drainage
    Reinstatement of drains 100 mm dia. per m run.

    Extra over junctions        .......... No.
      "    "     "              .......... No.
    Reinstatement of gulleys    .......... No.

10. Reinstatement of concrete path      per. sq. m .......

11. Daywork
    For works not envisaged in the Specification and the Bill
    of Quantities.
    Labour rate including all usual extras      ..........
    Plant plus percentage including fuel and transportation to
    and from the Site            ......................

12. The above lump sums  and rates will be deemed to satisfy all
    aspects of the Specification irrespective of the descriptions
    applied.
```

General References

Burland, J.B., Butler, F.G., and Dunican, P., 'The behaviour and design of large diameter bored piles in stiff clay', *Proceedings of Symposium on Large Bored Piles,* 1966.

Methods of calculating the ultimate bearing capacity of piles. Brom Sols – Soils 5, 1966.

Peck, R.B., Hanson, W.E., and Thorburn, T.H., *Foundation Engineering,* John Wiley, New York, 1967.

Steelwork Designers' Manual, Crosby Lockwood, 1972.

Properties and safe load tables. *Structural Steelwork Handbook,* BCSA.

Hodgkinson, A.J. (ed.), *AJ Handbook of Building Structure,* 2nd edn, Architectural Press, 1980.

Chapter 6

Examples of Foundation Design

6.1 Simple Bases

6.1.1 Column on Square Base

The average ground pressure is $\dfrac{W}{a^2}$

The thickness of the base must be so proportioned that the column does not punch through to the soil below, the base does not fail in shear or in bending. Assume W = 200 kN.   a = 1 m,   c = 300 mm on a soil where the allowable bearing pressure has been assessed at 214 kN/m². (2 T/sq.ft).

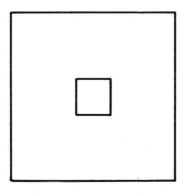

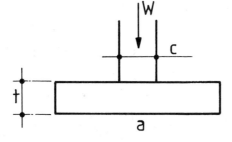

Punching shear

The load spreads down from the column into the shaded portion of area $(c + 3t)^2$ and generates a shear along the perimeter line length $4c + 12t$. The passive upload from the soil is given by $\dfrac{W}{a^2}(a^2 - (c + 3t)^2)$ on the area outside the shear perimeter so that the shear stress on this line is given by $\dfrac{\dfrac{W}{a^2}(a^2 - (c + 3t)^2)}{(4c + 12t)t}$

$$\frac{W}{a^2} = 200 \text{ kN/m}^2 \quad c = 300 \text{ mm.}$$

Assume that the shear stress is fs. N/mm².

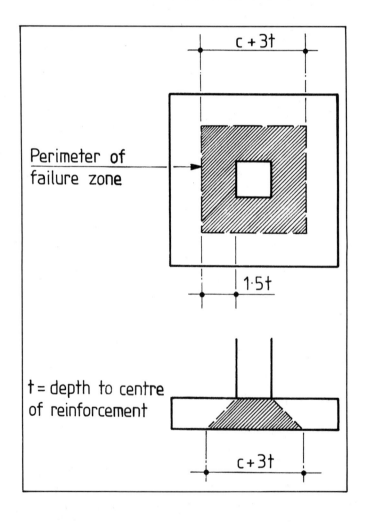

Perimeter of
failure zone

t = depth to centre
of reinforcement

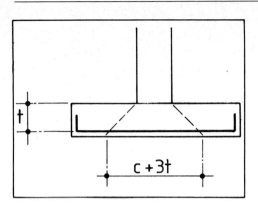

$$200,000 \left[1 - \frac{(300 + 3t)}{1000} \right]^2 = fs. (1200 + 12t)t$$

and this presents a quadratic equation from which the value of t is given in mm. This is the depth to the reinforcement so the overall thickness of the base becomes $t + \frac{1}{2}d + 50$ where d is the bar diameter and 50 mm is the cover required against the earth.

Bending shear

As in the punching shear the column spreads the load out directly to the soil over a distance 1·5t from the column face leaving the passive upload from the soil as

$$\left(\frac{a}{2} - \frac{c}{2} - 1\cdot5t \right) \times a \times \frac{W}{a^2}$$

Again if fs is the shear force

$$fs. a. t. = \left(\frac{a}{2} - \frac{c}{2} - 1\cdot5t \right) a \times \frac{W}{a^2}$$

$$\text{or } t = \frac{\left(\frac{a-c}{2} \right) \dfrac{W}{a^2}}{\dfrac{(fs + 1\cdot5W)}{a^2}}$$

Critical section
for shear

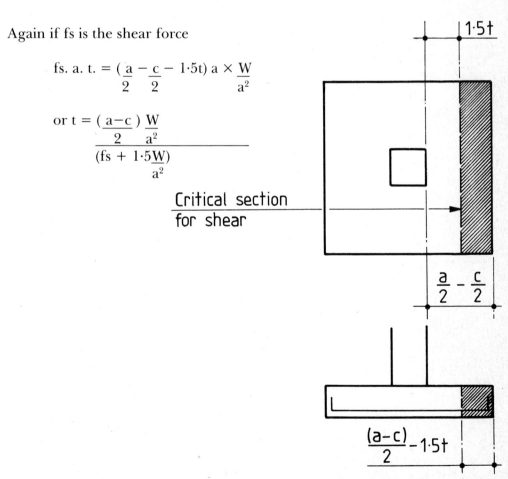

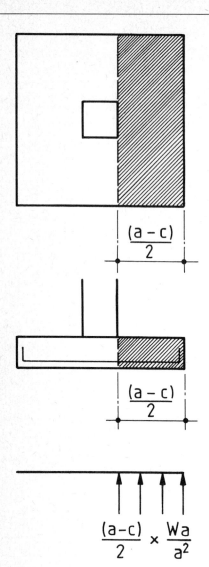

$$\frac{(a-c)}{2}$$

$$\frac{(a-c)}{2}$$

$$\frac{(a-c)}{2} \times \frac{Wa}{a^2}$$

Bending moment
The critical section for bending is at the face of the column and the passive load on the shaded area bends the base upwards from the right putting the reinforcement into tension.

$$\text{The bending moment is } \frac{(a-c)}{2} \frac{W}{a} \times \frac{(a-c)}{2} \frac{1}{2}$$

This is carried by the full width section 'a' of effective depth 't' and the amount of reinforcement is then calculated. In bases more than in other reinforced members except pile caps it is important to ensure that the diameter of reinforcement chosen is such that the bond length can be established. This may mean bending the bar well up the face of the base.

When reinforcing such bases, whether rectangular or square, it must be remembered that the same calculation must be made in each direction at right angles; the innocent might be tempted to assume bending of only half the total load in each direction.

There is an obvious concentration of load underneath the column and it is good practice to put ⅔ of the total reinforcement into the mid width of the base.

6.1.2   The Eccentric Load on a Rectangular Base

Eccentric loads can derive from two situations. The load itself may be out of centre of the base in two directions at right angles, or the central column may be subjected to an added bending moment.

The latter is the more general case and is now considered with load W and moment M. Assume the base is of sides a and b.

If the magnitude of M is such that the ground pressure at A remains positive then the problem is quite simple the average ground pressure being $\frac{W}{ab}$

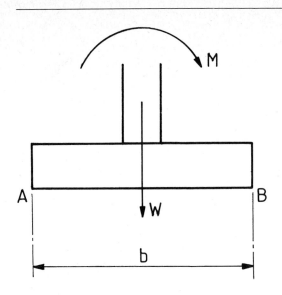

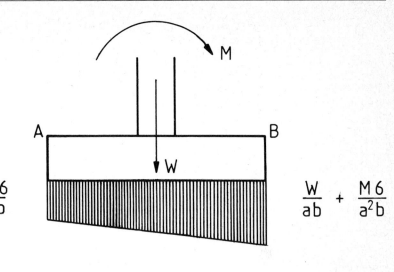

$$\frac{W}{ab} - \frac{M\,6}{a^2b} \qquad \frac{W}{ab} + \frac{M\,6}{a^2b}$$

and the bending moment ground pressure being the moment divided by the modulus of the plan shape of the base,

i.e. $\dfrac{M}{a^2b/6}$ negative at A positive at B

However, the value of M may be such that the answer at A. $\dfrac{W}{ab} - \dfrac{M6}{a^2b}$ may be negative. This means that the corner of the base A is effectively not inducing any passive pressure from the soil and the pressure diagram has changed to

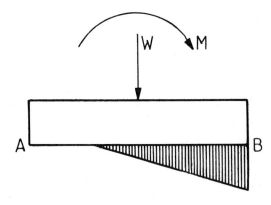

This would require a far more complicated calculation to arrive at the soil stress under the base. However, falling back on the Newtonian principle that reactions are equal and opposite, the W/M situation can be represented by

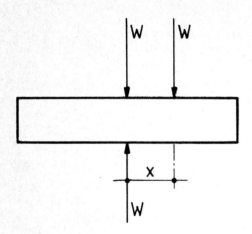

where x = $\dfrac{M}{W}$ and this in turn becomes

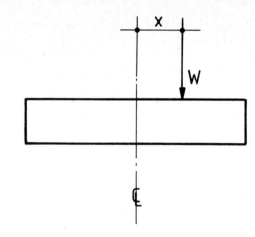

where x is described as the eccentricity about the base centre.

Now applying the equal and opposite theory, the soil pressure resultant must lie beneath W and in turn must be the centre of gravity of the triangle of soil pressure.

Now the pressure p is twice the average pressure which is $\dfrac{W}{3lb}$ and l is calculable as it is $\dfrac{(a-x)}{2}$

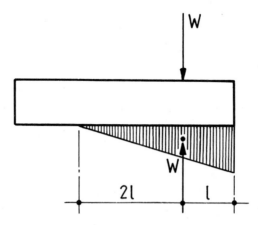

i.e. provided the eccentricity can be determined, the centre of gravity of the soil pressure is directly beneath it and this can often produce a very quick result when a number of base lengths are being investigated.

6.1.3   The Eccentric Load on a Trapezoid Base

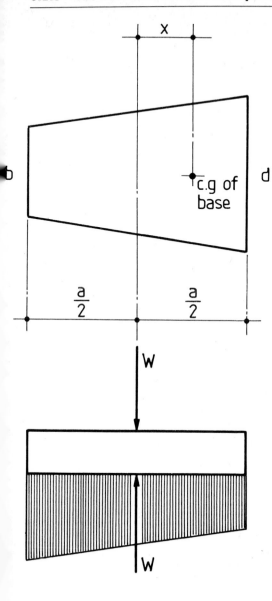

This case was considered in Chapter 3, where the proximity of the base to an existing building required a width greater at the building face to get a reduced ground pressure at that position.

Direct load W. Moment Nil.

The eccentricity is directly obtainable here as x, there being no applied moment. The bending moment about the centre of gravity of the base is Wx but again for the balance of upward passive and downward active loading the centre of gravity of the passive soil load will coincide with the applied load W.

The moduli of section of the trapezium are Z1 and Z2 where

$$Z1 = \frac{(b^2 + 4bd + d^2)}{(12(2b + d))}.a^2 \qquad Z2 = \frac{(b^2 + 4bd + d^2)}{(12(b + 2d))}.a^2$$

The ground pressures are

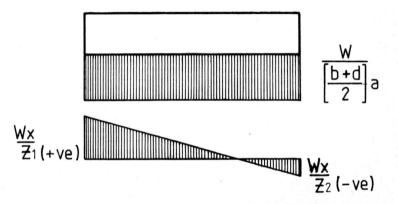

$$\frac{W}{\left[\dfrac{b+d}{2}\right]a}$$

$$\frac{Wx}{Z1}(+ve) \qquad \qquad \frac{Wx}{Z2}(-ve)$$

If by chance the load W was directly over the centre of gravity of the trapezium then there would be an average load of

$$\frac{W}{\left(\dfrac{b + d}{2}\right) \times a}$$

over the whole area.

If, however, the left hand side of the trapezium went negative in the above calculation, perhaps because a large moment of a clockwise direction is applied with the direct load, then the pressure diagram would be

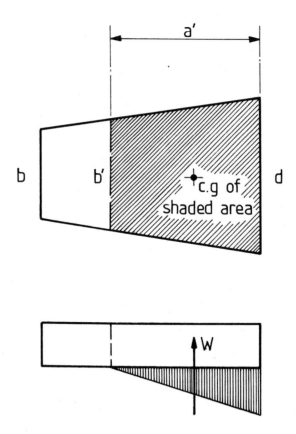

and the revised position of the aplied load W would now be directly over the centre of gravity of the shaded area, i.e. trapezium a'b'd. and the calculation would have to be repeated for the new trapezium with sides a',b',d

6.1.4   The L shaped Base

Probably a rarity, but occasionally employed in very difficult boundary conditions.

Assume that the area of the two legs equals A. The moduli about the Y axis equals Z_{y1}, Z_{y2}, and about the X axis equals Z_{x1}, Z_{x2}

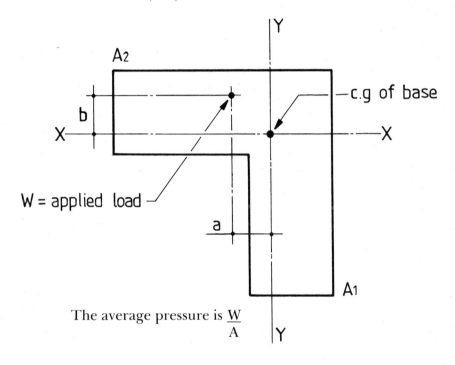

The average pressure is $\dfrac{W}{A}$

The eccentric moment about the XX axis is Wb.
The eccentric moment about the YY axis is Wa.
At the extreme edges

$$A2 \quad \text{pressure} = \frac{W}{A} + \frac{Wb}{Z_{x1}} + \frac{Wa}{Z_{y1}}$$

$$A1 \quad \text{pressure} = \frac{W}{A} + \frac{Wb}{Z_{x2}} - \frac{Wa}{Z_{y2}}$$

Provided the expression at A1 is still positive and the pressure at A2 does not exceed the allowable, the L shape is acceptable.
Rather than try to calculate the pressures for a condition where the expression at A1 became negative, it would be preferable to change the dimensions of the L.

6.1.5   The Circular Base

The circular base is likely to be found under the shaft of a circular elevated water tower or a gas holder.

In the former case there would be a high wind moment applied to the tank when it might be empty, thus producing an eccentricity in the circular base.

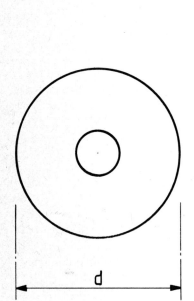

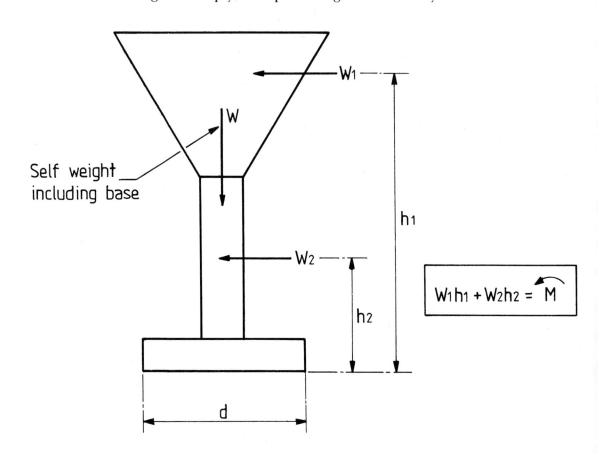

$$W_1 h_1 + W_2 h_2 = M$$

The direct load produces an average pressure $\dfrac{W}{\dfrac{\pi d^2}{4}}$

Assuming there is no part of the base not producing pressure on the ground the extreme edge pressures would on the windward side.

$$\frac{W}{\frac{\pi d^2}{4}} - \frac{M}{\frac{\pi d^3}{32}} \qquad \text{and on the leeward side} \qquad \frac{W}{\frac{\pi d^2}{4}} + \frac{M}{\frac{\pi d^3}{32}}$$

a simple calculation and obviously a base size is to be aimed at to provide the situation where both pressures are positive. The proximity of services might make this impossible as altering the shape, or the eccentricity could be such as to produce a negative pressure on one edge.

The loaded area would then be a segment of the circle

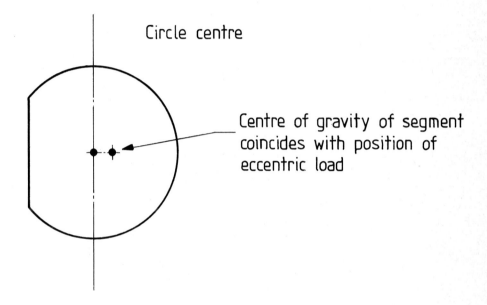

Circle centre

Centre of gravity of segment coincides with position of eccentric load

The moduli of the segment can be found from geometrical tables though obviously the problem is somewhat more complicated than the previous examples.

6.2   Column on the Edge of the Building

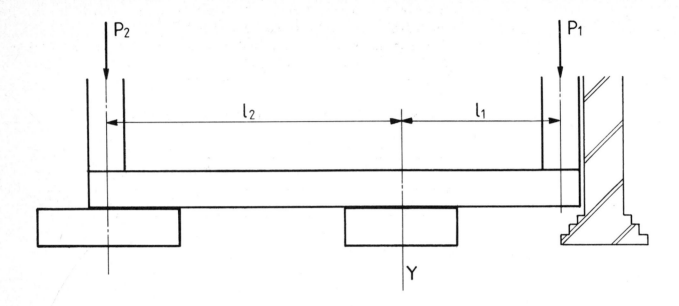

To avoid underpinning the adjacent foundation to accommodate a base for P_1 or, where P_1 is a short distance away from the adjacent property, to avoid any disturbance to the adjacent property, the tailing down beam method is often employed. For this to be a safe construction there must obviously be a load at P_2 such that $P_2 l_2$ is greater than $P_1 l_1$. Again, as the loads at P_2 and P_1 are made up of dead loads and superimposed loads a consideration must be made of the effect of P_1 receiving all its superimposed load while P_2 receives none, although the dimensions of the column layout usually produce floor panel sizes which make the condition unlikely.

If the weight of the base at P_2 is ignored, the reaction uplift at P_2 due to P_1 should be increased by 10% for the dead load and 25% for the superimposed load.

$$\text{i.e.} \frac{P_1 \text{ dead} \times l_1 \times 1 \cdot 1}{l_2} + \frac{P_1 \text{ live} \times l_1 \times 1 \cdot 25}{l_2} = P_2 \text{ dead}$$

There is the small effect of the weight of the tailing down beam to be added to the above.

Base Y is designed for

$$P_1 + \frac{P_1 l_1}{l_2} + \text{beam effect}$$

6.3 Foundation with a Number of Loads

Assume that four columns on a 4·5 m grid are to be founded on a gravel stratum the top of which is 2 m below ground level.

The excavation is to be excavated in trench form with a width of 900 mm and the beam foundation depth is 1550 mm.

The net ground pressure is not to exceed 300 kN/m².

The loading data and dimensions are shown below:

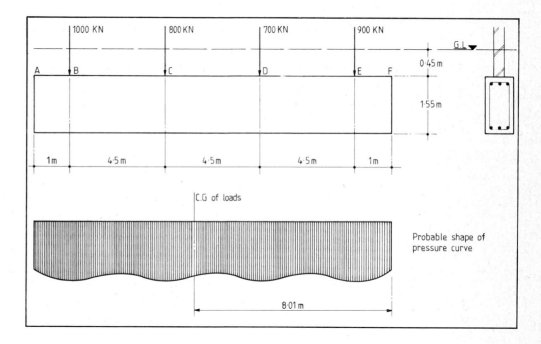

For the span depth ratio B to E of $\dfrac{13\cdot5}{1\cdot55}$ = 8.71 the beam can be assumed rigid and the pressure distribution a straight line.

First of all calculate the ccntre of gravity of the loads.

Taking moments about E

$$\frac{1000 \times 4\cdot5 \times 3 + 800 \times 4\cdot5 \times 2 + 700 \times 4\cdot5 \times 1}{1000 + 800 + 700 + 900} = 7\cdot01 \text{ m from E}$$

or 8·01 from F.

Then the eccentricity about the centre line of the base = 8·01 − 7·75 = 0·26 towards C from the centre.

This produces a moment of 3400 × ·26 = 884kNm.

$$\text{The ground pressure at A} = \frac{3400}{\cdot9 \times 15\cdot5} + \frac{884 \times 6}{\cdot9 \times 15\cdot5^2} = 268\cdot26\text{kN/m}^2$$

$$\text{The ground pressure at F} = \frac{3400}{\cdot9 \times 15\cdot5} - \frac{884 \times 6}{\cdot9 \times 15\cdot5^2} = 219\cdot20\text{kN/m}^2$$

The pressure diagram becomes

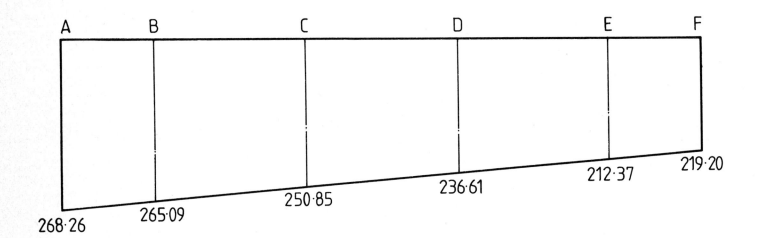

The shear force diagram for the beam foundations can then be drawn as follows:

$$\text{Pressure on base from A to B} = \frac{(265\cdot09 + 268\cdot26)}{2} \times 1 \times \cdot9 = 240.01$$

$$\text{B to C} = \frac{(265\cdot09 + 250\cdot85)}{2} \times 4\cdot5 \times \cdot9 = 1044.78$$

$$\text{C to D} = \frac{(250\cdot85 + 236\cdot61)}{2} \times 4\cdot5 \times \cdot9 = 987.11$$

$$\text{D to E} = \frac{(236\cdot61 + 222\cdot37)}{2} \times 4\cdot5 \times \cdot9 = 929.43$$

$$\text{E to F} = \frac{(222\cdot37 + 219\cdot20)}{2} \times 1 \times \cdot9 = 198.71$$

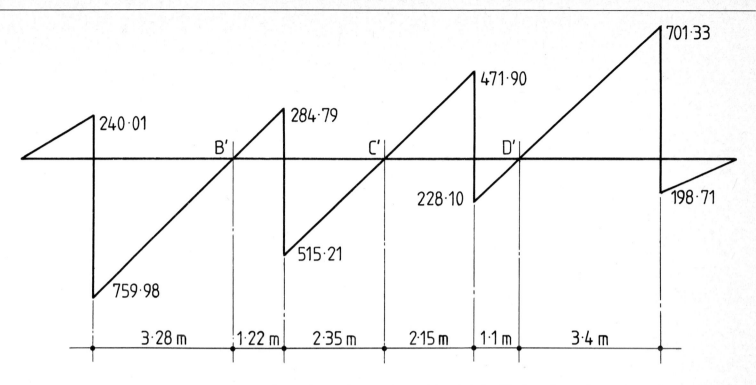

The points of zero shear B^1, C^1, D^1 will give the points of maximum bending moments in the bays BC, CD and DE and the bending moments at each of the lettered points can be calculated by summing the upward and downward moments at each position. It is easier to start from F and work from right to left. The pressure diagram is split into two parts – a rectangle and a triangle – which have easily identifiable centres of gravity.

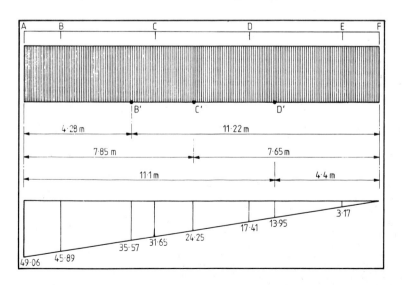

A typical calculation of b^1 is

$$35 \cdot 57 \times \frac{\cdot 9}{2} \times \frac{1}{3} x\ 11 \cdot 22^2 + 219 \cdot 2 \times \frac{\cdot 9}{2} \times 11 \cdot 22^2 - 900 \times 11.22$$
$$- 700 \times 5 \cdot 72 - 800 \times 1 \cdot 22 = -1087\ \text{kNm}.$$

and the full answers are

Bending mt at A $= 0$
B $= +121$
$B^1 = -1087$
C $= -911$
$C^1 = -911$
D $= -1504$
$D^1 = -1109$
E $= +100$
F $= 0$

where + ve means the bottom is in tension and −ve the top and the foundation bends thus

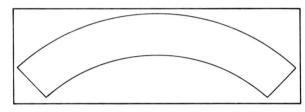

The bending moment diagram can be plotted thus:

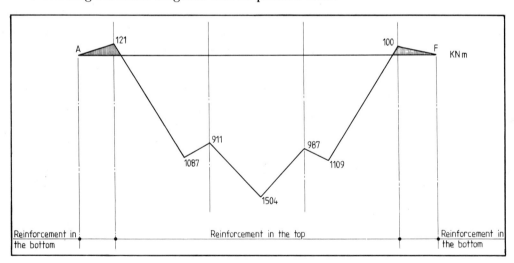

The worst bending moment is 1504 kNm, which could actually be carried by a beam 900 mm wide by only 520 mm deep, but the span depth ratio would have gone up from 8·71 to 26·0 and the moment of inertia of the lesser beam would have been only 0·04 of that used in the design. Obviously with such a lesser depth beam the above method could not have been used.

Where comparatively shallow beams have been used, e.g. span/depth ratio 16·0, approximate methods have been employed such as designing the beam for the load spread between the columns.

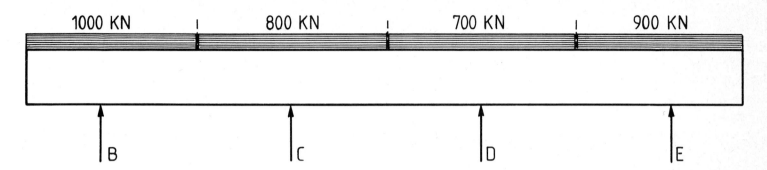

Obviously this cannot be correct as the analysis of the foundation, as a beam loaded in this way will not give support reactions equal to the 1000, etc., loads applied at those positions.

If the beam is flexible or the soil is clay, then there is no alternative but to employ a method which by iteration eventually produces ground pressures where the support reactions equal the applied loads or, put another way, the deflected shape of the soil is exactly that of the deflected shape of the beam. Such an example could not be illustrated here.

With the development in recent years of the finite element method and computers with enormous storage capacity, more direct methods of solution

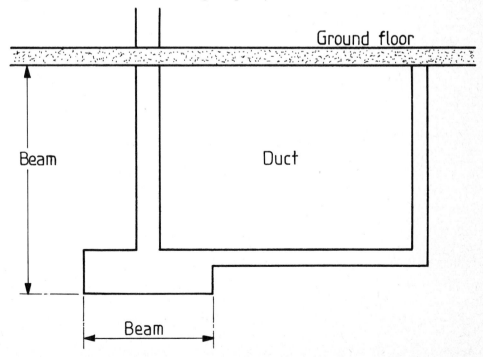

have been developed in which the common displacement profile and ground reactions at the common interface are determined in a single operation. The various methods are described in the Institution of Structural Engineers' State of the Art Report on *Structure–Soil Interaction.*

The foundation in the example was designed as a rectangular beam but could have been shaped alternatively as a tee beam. This shape is often the natural result of running a large service duct along the line of columns in a main walkway through a school or office block at ground floor.

6.4 Examples of Piling

6.4.1   Piled Foundation for a Two-Storey House

A house, as part of a large development, is to be erected in an area where the site investigation has given the following values for the clay:–

Plasticity index	35%
Clay fraction	30–60%

Cohesion increases from 50 at 1 m to 250 at 10 m kN/m².
The site requires filling exceeding 600 mm
Outer walls are 105 brick 100 block
Inner spine wall is 150 block.

It is not immediately obvious which wall is most heavily loaded

$$
\begin{array}{lll}
 & & \text{kN/m} \\
\text{At C roof load per m} & = \dfrac{(9 + \cdot3)}{2} \times 1\cdot75 \ = & 8\cdot40 \\
\text{1st floor per m} & = \dfrac{5}{2} \times 2\cdot1 \qquad = & 5\cdot25 \\
\text{Ground floor per m} & = \dfrac{5}{2} \times 3\cdot5 \qquad = & 8\cdot75 \\
\text{Wall less windows per m} & = 5\cdot4 \times \cdot8 \times 3\cdot2 \quad = & 13\cdot82 \\
 & & \overline{36\cdot22} \\
\text{At B roof load per m} & = \qquad = & 0 \\
\text{1st floor per m} & = \dfrac{9}{2} \times 2\cdot1 \qquad = & 9\cdot45 \\
\text{Ground floor per m} & = \dfrac{9}{2} \times 3\cdot5 \qquad = & 15\cdot75 \\
\text{Wall less doors per m} & = 5\cdot4 \times \cdot95 \times 1\cdot5 = & 7\cdot70 \\
 & & \overline{32\cdot90}
\end{array}
$$

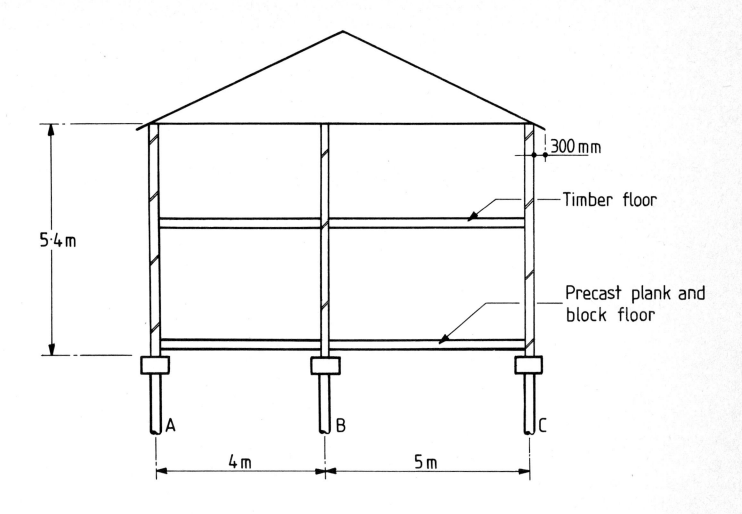

The actual position of piles will be related to the length of the walls involved and there may be cross-wall loads in the form of partitions or main walls carried on beams which are taken directly to the piles or via the connecting pile head beams. Assume that this raised the calculated loads to 45 kN/m and that the layout of the house is such that piles are placed at 2 m centres.

The pile will thus be required to carry the 90kN plus the connecting beam load of say $\cdot3 \times \cdot3 \times 24 \times 2 = 4\cdot4$ kN. a total of $94\cdot4$ kN.

Having regard to the site conditions the top 1·5 m of the pile must be disregarded as load bearing. The cohesion with depth diagram is

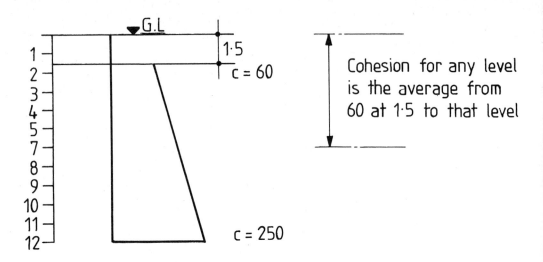

Assuming auger bored piles diameter 300 mm and an ultimate value of ·45c in adhesion and 9c in bearing at the base ultimate values can be calculated at 1 m intervals in the clay as follows measured from ground level.

The reason for the reduction from 1·0 to 0·45 in the adhesion is to allow for deterioration in the cohesion value due to combined effects of softening and swelling of the clay, seepage of water from fissures in the clay and the wet concrete and softening from the type of boring operation.

Table 6.1

Depth below GL	c kN/m²	Adhesion kN $\cdot45 \times \cdot3 \times \pi \times c \times (d - 1\cdot5)$	c kN/m²	Base capacity kN $9c \times 3^2 \times \pi/4$	Total kN
3	77	49	94	60	109
4	89	93	117	75	168
5	100	147	139	89	236
6	111	210	161	103	313

Using a factor of safety of 2·5 on the design load of 94·4 gives an ultimate design load of 236 kN so the piles would need to penetrate to a depth of 5 m below ground.

6.4.2   Large Bored Cylinder Pile

Consider the design of a pile to carry a working load of 2000 kN with a safety

factor of 2·5, i.e. an ultimate load of 5000 kN.
At 16 m below ground level the pile length is 2 m in the gravel and 12 m in the
clay. In the gravel the skin friction may be determined from the table.

Table 6.2

Density	N value	Ultimate skin friction kN/m²
Loose	4–10	10–20
Medium	10–30	20–50
Dense	30–50	50–100
Very dense	>50	100

In this case take a friction value of 30. In the clay the ultimate load capacity is
based on adhesion on the cylinder sides based on the average cohesion value
between 4 m and 16 m depth and the end bearing based on the cohesion at 16 m
depth.
The Site Investigation Report has provided the results of triaxial tests at
different depths in the clay which is overlain by 2 m of gravel and 2 m of newly
placed fill.
The plot of the triaxials is as follows.

The numbered dots represent the cohesion at that depth in that particular
numbered borehole and a straight line is then drawn along the median line
between the dots. This is then described as the 'average cohesion'.

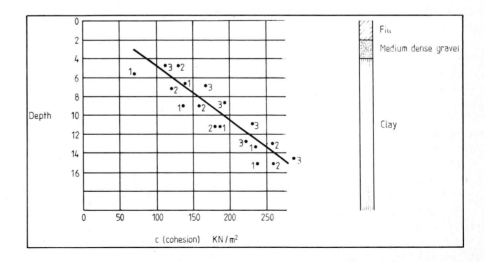

At 16 m depth. Ult. clay adhesion on a 1 m dia. pile

$$\pi \times 1 \times \frac{(300 + 75)}{2} \times 12 \times \cdot 45 \qquad\qquad = 3180$$

Ult. gravel friction do. $\pi \times 1 \times 2 \times 30$ $\qquad = \ \ 189$
End value in clay $= 9 \times \pi \dfrac{1^2}{4} \times 300$ $\qquad = 2120$

$\overline{}$
5489
$\overline{}$

This compares with the ultimate applied load of 5000 kN.
However the 2 m of new fill creates negative skin friction for which the Site Investigation Report recommends a down drag of 200 kN for the 2 m depth of fill.

$\qquad\qquad\qquad\qquad\qquad\qquad\qquad$ kN

The factored carrying capacity of the pile $= \dfrac{5489}{2\cdot 5} = \ \ 2195\cdot 6$

$\qquad\qquad\qquad\qquad\qquad$ Down drag $\qquad \underline{\ \ 200\cdot 0}$

$\qquad\qquad\qquad\qquad\qquad\qquad\qquad \underline{1995\cdot 6}$

which is near enough to the 2000 kN applied load.

6.4.3 Large Bored Pile with Under-ream

Consider the above problem solved by using an under-reamed pile 600 mm shaft 1800 mm under-ream and at a depth of 13 m average cohesion $\dfrac{(250 + 75)}{2} = 162$.

base cohesion 250 kN/m². The shaft adhesion is reduced by using a factor of 0·3 instead of 0·45 in view of the under-ream and 2 m is reduced in the adhesion length above the under-ream.

$\qquad\qquad\qquad\qquad\qquad\qquad\qquad\qquad\qquad\qquad$ kN

At 13m depth Ult. clay adhesion $\pi \times \cdot 6 \times 162 \times 8 \times \cdot 3 = \ \ \ \ 733$

$\qquad\qquad$ Ult. gravel friction $\pi \times \cdot 6 \times 2 \times 30 \qquad = \ \ \ \ 189$

$\qquad\qquad$ End value in clay $9 \times \pi \times \dfrac{1\cdot 8^2}{4} \times 250 \qquad = \ \ \underline{5726}$

$\qquad\qquad\qquad\qquad\qquad\qquad\qquad\qquad\qquad\qquad\quad 6648$
$\qquad\qquad\qquad\qquad\qquad\qquad\qquad\qquad\qquad\qquad\quad \overline{}$

Allowing for the negative fill load

$$\text{The factored capacity of the pile} = \frac{6648}{2\cdot5} = \overset{kN}{2659}$$

$$\text{Down drag} = \frac{200}{2459}$$

When using an under-ream and shaft the pile loads must also satisfy a criterion proposed by Burland, Butler and Dunican (Large Bored Piles symposium, 1966) that the working load should be less than ultimate shaft load + ⅓ ultimate base load, i.e. (733 + 189) + 1/3 (5726) = 2831. This can of course be scaled down to $2831 \times \frac{2000}{2659} = 2129$ which is near enough.

6.4.4 Pile Settlement

Note that in both bases 6.4.2 and 6.4.3 the piles should be taken to a depth sufficiently great to avoid an excessive settlement.
A curve proposed by Burland, Butler and Dunican shows the settlement relationship for piles in clay using the plate-bearing test.

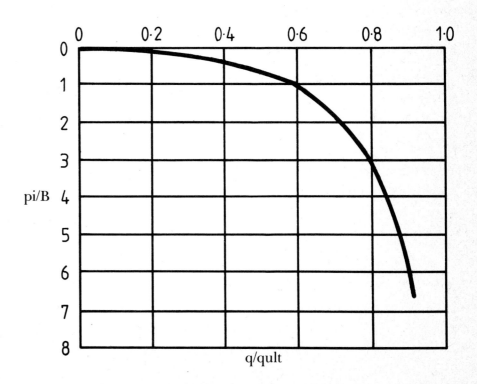

$$pi = \frac{\cdot6qB}{E_d}$$

where pi/B is <u>plate settlement</u>
                    plate diameter
q/qult = <u>plate bearing pressure</u>
                    ultimate bearing capacity of soil beneath plate
E_d = deformation modulus
K the bearing plate factor = $\cdot01 - 0\cdot02$ in hard clay.
The settlement can be calculated for a circular base pi $= \frac{\cdot6qB}{E_d}$

In 6.4.3 the load at working load on the base of the pile is $2000 - 733 - 189 =$
1078 causing a pressure in the base of $\frac{1078}{\pi\times1\cdot8^2/4}= 424$ kN/m²

A preliminary estimate of the elastic settlement is therefore =
          $\cdot02 \times 1\cdot8 \times 1000 \times 424/(9 \times 250) = 7$ mm (O.K. $\cdot02$ is conservative).

6.4.5 Bored Pile in Sand

Assume that the Site Investigation Report shows a dense sand with N values at 2 m (25) at 4 m (35) then consistently 50 from 6 m. On top of the sand is a 3 m deep fifty-year-old fill which will cause no negative friction. The design load is 400 kN.
Using equations of Chapter 2, unit skin friction is given by using the equation $\bar{K}_s \gamma \alpha \tan \delta$
where $\bar{K}_s$ and δ are factors suggested by Broms.
$\bar{K}_s$ is an earth pressure coefficient, δ is the angle of wall friction.

Table 6.3

Pile Material	δ	$\bar{K}_s$	
		low relative density	high relative density
Concrete	$\tfrac{3}{4}\phi$	1·0	2·0

The angle δ can be obtained from a curve proposed by Peck.
Assume a pile diameter of 350 mm and depth 10 m.

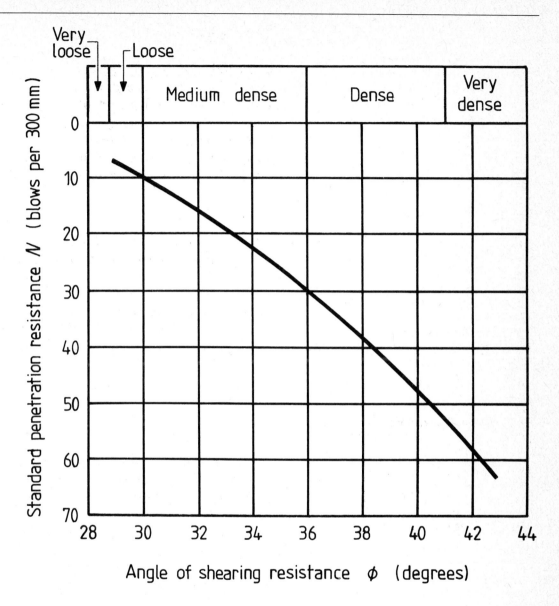

Angle of shearing resistance ϕ (degrees)

S.P.T.

In this case $\bar{K}_s = 1$ ϕ for 25 for 2m–4m 35° $\cdot 75 \times 35 = 26.2$
                                35 for 4m–6m 37° $\cdot 75 \times 37 = 27.8$
                                50 thereafter 41° $\cdot 75 \times 41 = 30.75$

Skin friction per unit at 2 m $= 1 \times 18 \times 2 \times \tan 26 \cdot 2\ = 17 \cdot 7$
Skin friction per unit at 4 m $= 1 \times 18 \times 4 \times \tan 27 \cdot 8\ = 38 \cdot 0$
Skin friction per unit at 7 m $= 1 \times 18 \times 7 \times \tan 30 \cdot 75 = 75 \cdot 0$

At the toe reduce to $41° \times \cdot85$ to allow for the disturbance of the boring tool, i.e. $\varphi = \cdot35°$.

From the previous deep foundation Nq bearing factors graph $\phi = 35°$
$D/B = 22$ $Nq = 50$.

$$
\begin{array}{llr}
 & & \text{Kn} \\
\text{Skin friction over 2 m} = & 2 \times 17\cdot7 \times \pi \times \cdot35 & = \quad 39\cdot0 \\
2\text{ m--}4\text{ m} = & 2 \times \dfrac{(17\cdot7 + 38)}{2} \times \pi \times \cdot35 = & 61\cdot2 \\
4\text{ m--}7\text{ m} = & 3 \times \dfrac{(38 + 75)}{2} \times \pi \times \cdot35 & = \quad 186\cdot4 \\
\text{Base resistance} & = \dfrac{\pi \times \cdot35^2}{4} \times 18 \times 50 \times 7 & = \underline{\ 606\cdot1} \\
 & & \underline{892\cdot7}
\end{array}
$$

This would give a factor of safety of $2\cdot23$ and, as a factor of $2\cdot5$ would be desirable, increase the pile length to 11 m.

6.5 Dispersion of Column Loads

The dispersion of the foundation load has been dealt with earlier but it is also necessary to disperse the column load into the upper half of the foundation. The reinforced concrete column does not constitute a problem as there is material continuity between the column and the base.
Sometimes a different quality concrete is used in the column from that in the foundation and it might at first appear that the foundation concrete could be overstressed. However, the foundation is usually much bigger than the column and the concrete immediately below the column is constrained in the manner of the sample in the triaxial text so that a higher stress than the usual ¼ cube strength can be taken. A formula which links the cube strength uf_c of the column with the foundation concrete stress f_b is

$$ f_b = \frac{uf_c}{3} \sqrt[3]{\frac{A1}{A0}} \text{ where A1 is the column area} $$
and A0 is the foundation area

e.g. if $uf_c = 25\text{N/mm}^2$ $A1 = 300^2\text{mm}^2$ $A0 = 1000^2\text{mm}^2$
then $f_b = \dfrac{25}{3} \sqrt[3]{\dfrac{9}{100}} = 3\cdot73 \text{ N/mm}^2$

i.e., the cube strength of the foundation must not be less than

$\dfrac{3\cdot73}{\cdot4} = 9\cdot4 \text{ N/mm}^2$. It is of course unlikely that it would be less than 15N/mm^2.

In both the case of the column and the foundation, the bearing stress must not be greater than ·4 × cube strength of the respective concretes.

When the load is applied to the steel base plate of a stanchion, the whole load passes to the foundation via the plate and as this sets up tensile stresses horizontally it is advisable to place a 100 mm sq. reinforcing mesh parallel to the surface and about 75 mm from it as the bearing stress approaches ⅓ cube strength.

When the type of cement used allows the concrete to go on gaining strength with time, it is permissible to increase the allowable stress, e.g. at 6 months by 20%. If the structure is not going to convey the design load to the foundation before 6 months, then the foundation concrete cube strength can be enhanced by 20% for design purposes.

It is usual in a steel base to size the steel base so that the eccentricity of moment over direct load lies within the base plate.

Example: a steel base plate 600 mm × 300 m receives a direct load of 250 kN and a moment of 100 kNm from a stanchion.

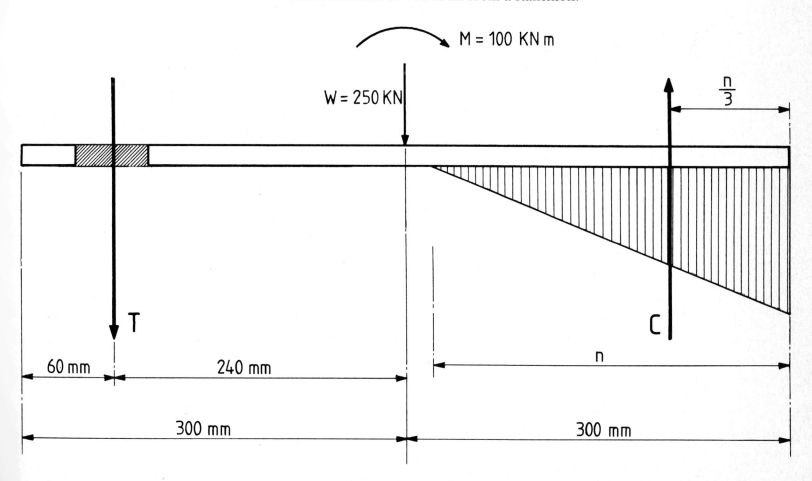

Check the adequacy of the base plate size and the size of the holding down bolts. The foundation concrete is of cube strength 20N/mm².

c = concrete bending stress compression force at C.G. of triangle

n = depth from compression edge to the neutral axis

By moments about T. $C \times (540 - \dfrac{n}{3}) = 250 \times 240 + 100 \times 1000$

$$= 16 \times 10^4$$

where C is in kN

By statics $T = C - 250$.

The allowable stress in the bolts is 120 N/mm² on the area at the base of the thread.

If the maximum concrete stress in bending is $\dfrac{20}{3}$ N/mm² then

$$C = \cdot 5 \,.\, n \,.\, 300 \,.\, \frac{20}{3} \,.\, \frac{1}{1000} = n$$

and $(540 - \dfrac{n}{3}) n = 16 \times 10^4$ or $0 = n^2 - 1620n + 480\,000$

Whence $n = \dfrac{1620 \pm \sqrt{1620 - 4 \times 480\,000}}{2} = 390\cdot5$ mm

Then $C = \cdot5 \times 390\cdot5 \times 300 \times \dfrac{20}{3} \times \dfrac{1}{1000} = 390\cdot5$ kN

and $T = 390\cdot5 - 250 = 140\cdot5$ kN

with 2 bolts area A at the bottom of the thread

$A = \dfrac{140\cdot5 \times 1000}{2 \times 120} = 585$ mm²

From the BCSA Handbook, 5858 mm² tensile stress area represents at 36 mm diameter bolt.

Chapter 7

Specifications

Whenever a project is sent out to tender it is preferable that a copy of the Site Investigation Report is included. It may well be that there are certain sections of the Report which are not directly relevant in which case a photostat of the remaining section can be included.

There is a common practice of advising the contractor that a copy of the report may be seen at the office of the engineer. This is most unsatisfactory as the cost of a visit by the contractor's estimator will be many times greater than the cost of a copy of the report and it is too much to expect the estimator to make all the necessary notes.

7.1   Mass Bases, Reinforced Pads and Trench Foundations and Pile Caps

Excavation specification

1. Bottoms of all excavations are to be taken down to the levels required by the drawings or other instructions, and are to be to the satisfaction of the Architect and Engineer and of the Local Authority.

2. All excavated material not required for backfilling shall be removed to a tip, to be provided by the Contractor.

3. The sides of the excavations shall be properly supported and retained by good sound timbering (or by other satisfactory methods) wherever necessary. All such timbering shall be carefully removed as the excavations are filled in. The removal of the timberings as aforementioned shall be done in such a manner as not to endanger the works and shall not relieve the Contractor of the responsibility for ensuring the stability of the Works.

4. The bottoms of all excavations shall be carefully trimmed and finished to the specified levels and all loose material removed.

5. Wherever the excavated surface has been cut up or has softened under the action of ponded water or has been broken up by any other cause the Contractor shall at his own expense excavate and remove soil down to a solid formation and backfill with concrete or fill as specified elsewhere properly consolidated up to the specified level.

6. If bad ground, cavities or soft spots are met with in any part of the excavation the Contractor shall excavate to a solid foundation and fill up to the specified level with fill or concrete as directed by the Architect and Engineer.

7. Should the Contractor excavate anywhere to a greater size (other than for working space) or depth shown on the working drawings or should the sides of the excavations cave in anywhere the Contractor shall at his own expense fill and tightly pack the excess spaces with concrete or other approved material.

8. The Contractor shall ensure that the formations are not damaged by weathering and concrete or fill shall be placed in the same day the excavation has taken place unless the formation is blinded with concrete or otherwise suitably protected from damage. A layer of plain concrete 1:8 mix 100 mm thick shall be laid on the bottom of prepared formations under concrete bases or strip footings when complete filling with concrete is not carried out on the day of the excavation.

9. The Engineer shall be informed before any concrete or hardcore is placed and shall be given the opportunity of inspecting and approving the bottoms of all excavations.

10. The Contractor shall make provision for and deal with all water which may find its way into the works from any source whatsoever and shall excavate sumps, cut drains, provide and work pumps and provide and work all necessary materials plant and apparatus for dealing with any water which may find its way into the excavations.

11. The Contractor shall not pump or otherwise put water directly into any drain.

12. Where reinforcement for concrete construction is to be placed on a blinding layer of plain concrete 1:8 mix 75 mm shall be laid to receive the reinforcement.

7.2 Information to be Given When Inviting Piling Tenders

The following is a checklist of the main items required; these should be prepared in detail by an experienced engineer.

1. Site plan with particulars of access.
2. Conditions of site when piling contractor starts work.
3. Responsibility for maintenance of surface during his operations.
4. Responsibility for setting-out, disposal of spoil, cutting away excess concrete deposited, removal of obstructions and provision of water and other facilities.

5. All factual information on soil investigation (not recommendations).

6. Any restrictions on working, e.g. noise or vibration, or limitations on headroom.

7. Information about underground obstructions or services on the site and adjoining properties.

8. Proposed layout of piling.

9. Working load on each pile. Settlement requirements of the structure.

10. Extent of piling contractor's responsibility for design. Necessary technical specification clauses.

11. Insurances required to be provided by piling contractor.

12. Form of main contract and piling subcontract or other general conditions of piling contract, as appropriate.

13. Dates for tender, start on site and completion of work.

14. Provision to be made for testing piles.

15. Any contingency sum to be allowed.

16. List of the different rates to be provided in the tender (assuming that a complete bill of quantities is not issued).

7.3 Notes for Guidance in Piling Inquiries and Administering Piling Contracts

1. Preliminary test piles. Whenever possible a preliminary test pile or test piles should be installed to check the pile design. These should be constructed under the closest supervision in an area where the soil conditions are known and tested to a specified load of not less than twice the working load.

2. Tests on working piles. Working piles for testing are selected at random by the engineer and should be tested to 1½ times the working load.

3. Factor of safety. Pile design should ensure that (a) an adequate factor of safety is provided against reaching ultimate load of the pile or pile group, (b) required load settlement characteristics are achieved at and near to design working load.

4. Negative skin friction. The usual method of providing for this is to calculate working load from ultimate bearing capacity using the factor of safety and then to add a net allowance for negative skin friction.

5. Concrete. Concrete for piles placed in the dry should contain not less than 300 Kg/m³ of cement and when placed under water by tremie tube a minimum cement content of 400 kg/m³ should be employed.

6. Concrete test cubes. Opinions vary as to the number of test cubes which should be required on a piling contract, but it is suggested that four cubes be taken for every 50m³ of concrete used. The anticipated number, of test cubes should always be included as a measured item in the bill of quantities.

7. Heave. The acceptable amount of heave depends upon whether piles are designed to carry the majority of their load by shaft friction. In cases where the pile carries the majority of its load in end bearing, heave may be reduced by pre-boring or, alternatively, the contractor may elect to redrive piles where this is a practicable solution. Particular measures required will vary with each site and those adopted should be a matter for discussion and agreement with the piling specialist concerned.

8. Tolerances and cut-offs. When deep cut-offs are involved or long temporary casings have to be used in construction of piles, it is not possible to estimate the amount of concrete to form the finished level within normal tolerances. In such cases the tolerance should be a matter for discussion and agreement with the piling specialist concerned.

9. Data sheet. A piling inquiry data sheet should be enclosed with all piling inquiries.

    When piles are designed by the client, the inquiry data sheet will give all details including pile diameters, length or penetrations required, reinforcement and concrete specification.

    When piles are to be designed by the piling specialist or when alternatives are permitted, the inquiry data sheet will state required factor of safety (if different from that given in the specification), acceptable settlement of individual piles under test at working load and any other basic requirements which must be fulfilled by alternative pile designs. The inquiry data sheet will entail any variations from the standard specification which are required by the client.

10. *In situ* soil testing. Where *in situ* testing is required, the type and anticipated number of tests should be indicated in the inquiry data sheet and included as a measured item in the bill of quantities.

7.4   Information to be Given when Inviting Piling Tenders

The following clauses are taken from Specification for cast-in-place piling, produced by the Federation of Piling Specialists (May 1971). They are given for general information, but the exact clauses to be used on a particular job must be decided by the engineer responsible.

Design

1. General design of piles shall be in accordance with the British Standard Code of Practice for Foundations CP 2004.
2. Average compression stress in concrete piles under working load shall not exceed 25% of the specified works cube strength at 28 days calculated on the total cross-sectional area of the pile shaft. Where the casing of the pile is permanent, the allowable compressive stress may be increased.
3. Piles shall normally be designed to carry all compression loads in the concrete on the cross-sectional area of the nominal diameter.
4. In the case of piles required to act in tension or bending, the stresses in the reinforcement shall be in accordance with CP 114.
5. Ultimate bearing capacity of a pile shall be taken as defined in CP 2004.
6. The factor of safety shall be taken as defined in CP 2004 as the ratio of the ultimate bearing capacity to the working load. Piles shall be designed to provide a factor of safety of not less than 2.
7. The cover of all reinforcement, where used, including binding wires, shall not be less than 40 mm.
8. The piles shall be designed to carry the working loads on the drawings and, in addition, allowance shall be made for stated negative skin friction loads.

Materials

9. Cement shall be ordinary or rapid hardening Portland cement complying with BS 12 or sulphate resisting cement complying with BS 4027.
10. Aggregate shall comply with BS 882.
11. Clean water free from acids and other impurities and in accordance with BS 3184 shall be used in the works.
12. All steel shall be in accordance with the appropriate British Standard unless otherwise agreed.

13. The slump of the concrete shall normally be in accordance with the following standards:

Piling mix	Slump minimum	Range	Typical conditions of use
A	75 mm	75-125 mm	Poured into water-free unlined bore. Widely spaced reinforcement leaving ample room for free movement between bars
B	100 mm	100-175 mm	Where reinforcement is not spaced widely enough to give free movement between bars. Where cut-off level of concrete is within casing. Where pile diameter is less than 600 mm
C	150 mm	150 mm or greater	Where concrete is to be placed by tremic under water or drilling mud

14. Any additive used in the concrete must be stated.
15. Ready-mixed concrete may be used and shall comply with BS 1926.
16. Test cubes shall be prepared and tested in accordance with BS 1881.

Driven Piles

17. Piles shall be installed in such sequence that their construction does not damage any piles already constructed.
18. Adequate measures shall be taken to overcome any detrimental effect of ground heave on the piles. When required by the engineer, levels shall be taken to determine the amount of any pile movement resulting from the driving process.

19. When a significant change of driving characteristics is noted, a record shall be taken of the driving resistance over the full length of the next adjacent pile (measured as blows per 250 mm penetration).

20. In the case of end bearing piles, the final set of each pile shall be recorded either as the penetration in millimetres per 10 blows or as the number of blows required to produce a penetration of 25 mm.

21. The temporary casing shall be dry after driving and before concreting commences.

22. Where cut-off level is less than 1·5 m below working level concrete shall be cast to a minimum of 150 mm above cut-off level. For each additional 0·3 m below working level of the cut-off level an additional tolerance of 100 mm will be allowed. Cut-off shall be a maximum of 3·0 m below working level. (See also clause 35.)

Bored Piles

23. A minimum length of 1·0 m of temporary casing shall be inserted in every borehole unless otherwise agreed.

24. When boring through non-cohesive or very soft cohesive strata liable to collapse, temporary casing or another suitable technique shall be used to stabilise the hole. Temporary casing when used shall extend a sufficient depth below such strata adequately to seal off the unstable material.

25. In dry non-cohesive strata water may be used to assist the advancement of the boring.

26. When subsoil water which cannot be sealed off is encountered the water in the bore shall be maintained above the standing level of the subsoil water.

27. When it is proposed to use a prepared drilling mud suspension the engineer must be advised.

28. When under-reaming of the bore is carried out the slope of the under-ream must be a minimum of 55 deg to the horizontal.

29. When it is not practicable to exclude groundwater from the finished bore the concrete shall be placed in a tremic tube.

30. Where cut-off level is less than 1·5 m below working level concrete shall be cast to a minimum of 150 mm above cut-off level. For each additional 0·3 m below working level of the cut-off level an additional tolerance of 50 mm will be allowed. (See also clause 35.)

31. When concrete is placed by tremic tube the concrete shall be cast to piling platform level or to a minimum of 1·0 m above cut-off level with a tolerance from 1·0 m to 2·0 m.

General

32. Piles shall be constructed within the following normal tolerances:
    In plan   75 mm in any direction at piling level
    Verticality   1 in 75
    Raking up to 1:6   1 in 25

33. Each batch of concrete in a pile shaft shall be placed before the previous batch has lost its workability. Removal of temporary casings must be completed before concrete within casing loses its workability.

34. In cold weather ice and snow shall be excluded from the material used in the manufacture of concrete for piles. Aggregate must not be heated to more than 38 deg C and the concrete when placed must have a minimum temperature of 5 deg C. The tops of the piles must be protected immediately casting is completed.

35. When concreting dry pile holes through water-bearing strata the concrete must always be cast to a minimum of 0·3 m above the standing level of the subsoil water unless all water-bearing strata are effectively sealed off by permanent casing and this level of 0·3 m above standing water level shall be regarded as cut-off level for the purpose of calculating tolerances as defined in clause 30 and clause 32.

36. Where concrete is not brought to piling platform level the empty pile holes shall be backfilled.

37. Safety procedures during piling operations shall comply with the recommendations of BS CP 2011[6] where applicable.

38. The following records shall be kept of every pile:
    pile number
    piling platform level related to datum
    nominal shaft/base diameter
    date driven or bored
    date concreted
    depth from piling platform level to toe
    depth from piling platform level to cut-off level
    depth from piling platform level to top of concrete
    final set (for driven piles), weight and drop of hammer
    length of permanent casing
    details of any obstructions encountered and obstruction time.

Chapter 8

Site Supervision

Supervision of site works is the responsibility of the contractor and requires continuing presence but the architect or engineer have a lesser responsibility to see that the standard of workmanship defined in their specification is maintained. This is usually achieved by going through the specification with the contractor (it is surprising how often the man on site has not even read it), discussing points of difficulty and contention with him and making such visits to the Site as the pace of construction demands, but, remember, the construction is being monitored not supervised.

8.1 Points to be Checked

(a) Excavation carried out to drawing dimensions, if in excess make sure proper filling is carried out.

(b) In trench-fill type construction, check that the machine has not left oversize teeth marks or crumbs at the bottom of the trench. Check that the trench has reasonably vertical sides, i.e. no greater than 1 in 75 slopes.

(c) In this type of construction if polythene, clay board or expanded polystyrene are being used on the trench sides, see that the materials have been placed so that the concrete cannot get behind them.

(d) It is the contractor's responsibility not to kill any of his staff, but if the architect or engineer consider that any excavation sides are unsound or liable to fall with the next heavy rain shower, they would be well advised to give their opinion in writing. The quantity surveyor will have measured planking and strutting but the contractor may take the risk and save that cost. Likewise, see that the contractor is not likely to destroy the adjacent property.

(e) Check that the conditions at foundation formation level are compatible with the site investigation data and make any depth or width variations to the foundations if necessary. Also check the standing water level as soon as an excavation is available.

(f)   Establish that there are permanent levelling reference points and grid line positions and ensure that the contractor carries out checks at intervals during construction. The levelling point is no use if it is ground subject to frost heave, swelling, etc.

(g)   If a clay bottom, check with pocket penetrometer against cohesion in the Site Investigation Report. If gravel or sand check by forking and ramming a crowbar into the surface.

(h)   If reinforced concrete members are to be constructed, ensure that the formation is blinded and the reinforcing cage given correct bottom and side cover as per the drawings. Do not allow unprotected sides of the excavation unless they are very stable, then shutter the concrete member or, if positively stable, cast the concrete into the excavated face allowing another 40 mm or so cover to the reinforcement.

(i)   If formwork is employed, then check that it is to line and level and sufficiently robust.

(j)   The concrete materials, if the concrete is mixed on site, should be stored in clean dry conditions with the cement preferably in a silo, otherwise in a shed.

Batching devices should be checked for accuracy, whether by gauge boxes or weight indicator, and aggregates should be given the opportunity to drain before being used.

(k)   If ready-mixed concrete is used, ensure that it is agitated until placed at site.

(l)   In either method of concreting, four cubes should be taken at the point of placing and when 24 hours old removed to a storage water tank and kept at about 60°F.

(m)   Generally, if the contractor makes tea as soon as the architect arrives, assume that he wishes to avoid an immediate inspection.

(n)   Remember that the most rewarding site visit is the one which is unannounced.

Preferred Reading

1. *Foundation Design and Construction,* M.J. Tomlinson, Pitman, revised edn 1980.

   The Author considers this book, first published in 1963 and several times revised and reprinted, to be the most comprehensive and yet easily read book available. It goes beyond the scope of this publication in dealing with retaining walls, earthworks and civil engineering works, but its appeal is that it is written by an engineer who spent many years with an international contractor so that all aspects are considered from a practical background.

2. *Physical Geography for Schools,* B. Smith, A. & C. Black, 1911.

   Despite the apparent elementary title of this book, first published in 1911 and now probably only found in libraries, it is in the Author's opinion one of the most readable books on general geology dealing with the composition of the land, earth movements, volcanoes, earthquakes and the history of the formation of the valleys, mountains and lakes as we know them today.

3. *A Geology for Engineers,* F.G.H. Blyth, Edward Arnold, 1945.

   Used by a number of universities as a geology course reference, this work takes the content of the above described book to a more advanced stage and studies rocks and minerals. There are chapters dealing with geological maps and the geology of water supply, reservoir and dam sites, cuttings and tunnels. Appendices cover soil mechanics and geophysical prospecting and engineering problems.

4. BS CP 5930 : 1981 Site investigations, BSI.

As well as dealing at length with all aspects of site investigations, BS CP 5930 contains valuable appendices covering the obtaining of information and the nature and occurrence of rocks and soils.

5. *Ground Engineering.*

This publication, issued eight times a year, provides a valuable means of keeping abreast of new techniques of construction and developments in soils engineering and describes important foundation works in progress.

Index